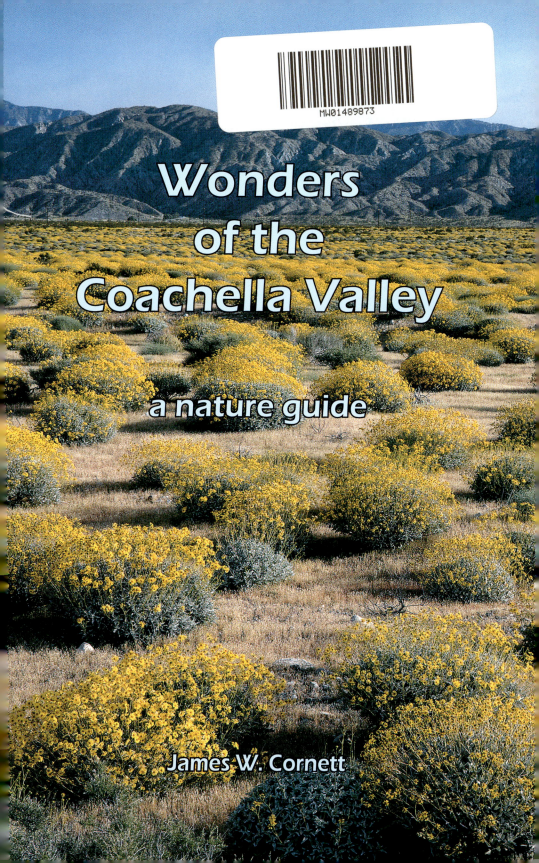

Wonders
of the
Coachella Valley

a nature guide

James W. Cornett

The following individuals reviewed the manuscript for technical accuracy: Harry Quinn (geology, anthropology), George Meyer (geology), Carl Garzynski (meteorology), Katie Barrows (botany), Kurt Leuschner (zoology), and Cameron Barrows (Coachella Valley Preserve). Jan Belknap and my wife, Terry Cornett, read the manuscript for consistency and clarity and helped transform my thoughts into words. My sincere appreciation is extended to each of these persons.

Special thanks to Jamie Lee Pricer and the Desert Sun for permission to reprint my series of articles that appeared in Desert Magazine. I also wish to express my gratitude to the Agua Caliente Band of Cahuilla Indians for access to the Indian Canyons Tribal Park.

Cover Photograph
Willis Palms, Coachella Valley Preserve

Inside Face Photograph
Brittlebush (*Encelia farinosa*) in bloom,
northwestern Coachella Valley

Photographs by the author

Copyright © 2008 by James W. Cornett
All rights reserved.

No part of this book may be reproduced by any means,
or transmitted or translated into a machine language,
without the written permission of the author.
International Standard Book Number (ISBN): 0-937794-02-3

P.O. Box 846
Palm Springs California 92263
Telephone 760-320-2774
Fax 760-320-6182

To my wife Terry

Contents

Valley Overview

10 Best Trips

How To Learn More

Dunes against the
Santa Rosa Mountains
in La Quinta

San Gorgonio Peak
11,499 Feet

San Bernardino Mountains

Yucca Valley

Morongo Valley

Highway 62

10

Desert Hot Springs

Pierson Blvd.

San Gorgonio Pass

Whitewater Canyon Road

Interstate 10

← To Los Angeles

Palm Dr.

Indian Ave.

Varner Road

Snow Creek Road

Highway 111

Tramway Road

2

Thousand Palms

Palm Springs

Mt. San Jacinto
10,804 Feet

3

Date Palm Dr.

Bob Hope Drive

Mt. San Jacinto
State Park

Cathedral City

Rancho Mirage

Palm Desert

Idyllwild

1

Palm Canyon

Palms to Pines Hwy.

Visitor Center

Palm Canyon Trail

10 Best Driving Trips

1 Palm Canyon
2 Palm Springs
   Aerial Tramway
3 Tahquitz Canyon
4 Palms to Pines Highway
5 Thousand Palm Oasis
6 Salton Sea
7 Painted Canyon
8 Joshua Tree
   National Park
9 Berdoo Canyon
10 Big Morongo
   Canyon Preserve

Highway 74

Highway 74

Pinyon

4

Santa Rosa and San Jacinto
Mountains National Monument

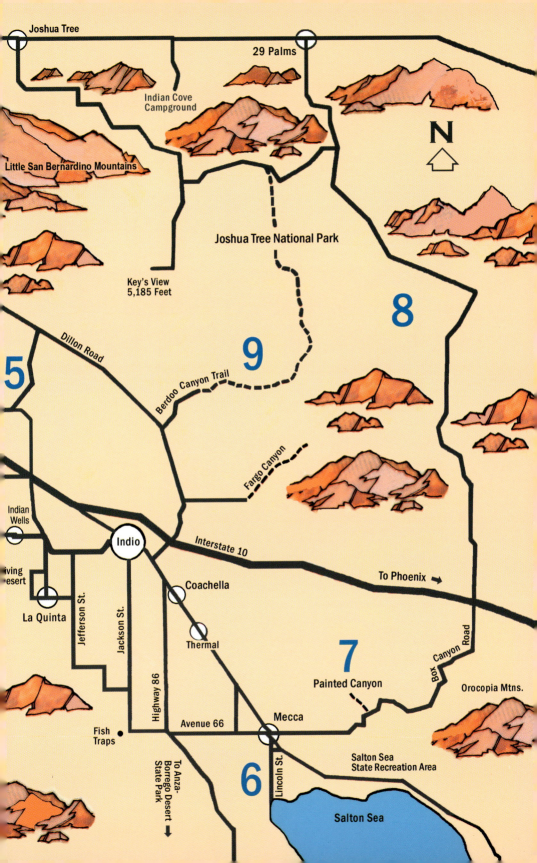

Welcome To The Coachella Valley

You have entered a region of amazing natural diversity. In a one-hour drive, a person can stand at the base of the steepest mountain in North America, stroll through the world's largest palm oasis, gaze upon rippled sand dunes, drive across the San Andreas Fault, and walk along the shore of California's biggest lake. There are few places in the world that contain so much landscape diversity in such close proximity.

Fortunately, for tourists and residents alike, most of this spectacular scenery is forever protected in eight preserves open to the public: Joshua Tree National Park, Santa Rosa and San Jacinto Mountains National Monument, San Jacinto Mountains State Park, Coachella Valley Preserves, Mission Creek Preserve, Indian Canyons Tribal Park, Salton Sea State Recreation Area, and Anza-Borrego Desert State Park. There is no other area in the country that has received so much protection from so many public and private agencies.

The first few sections of this book provide information on the natural environments of Palm Springs and the larger Coachella Valley in which Palm Springs is located. The section on geology briefly describes the valley's major landscape features, how they came to be and their role in creating a desert environment. The next section describes the valley's climate. Together, geology and climate have played a major role in determining the plant forms and species that exist in the Coachella Valley and the adaptations that allow them to survive in a desert environment. The following section describes the surprising diversity of wildlife and how individual species have adapted to the rigors of a region where rainfall is scant and hot summer temperatures are oppressive. The last introductory section describes the original human occupants of the region and how they made a living in a desert environment where resources seem in such short supply.

An early spring storm brings water to
sand verbena and brown-eyed primrose
in the central Coachella Valley

Most of the text consists of articles originally written for Desert Magazine, a monthly insert publication of the Desert Sun newspaper. These articles describe, in my opinion, the ten best places to visit while in the Coachella Valley. I have attempted to select a diverse array of destinations. For example, one must not miss Palm Canyon, the world's largest desert palm oasis, or the Palm Springs Aerial Tramway, the steepest tram ride in North America.

Each of the ten destinations can be visited in a single day's drive or less. In fact, the majority can be visited in less than half a day. Although this is not a hiking guide, hiking opportunities are readily available at each destination. With the exception of Tahquitz Canyon, all the destinations can be experienced without ever leaving the car (though I hope you will). Driving up Berdoo Canyon requires a high-clearance vehicle, ideally with all-wheel drive. Painted Canyon can usually be accessed with a street vehicle but to avoid bogging down in sand, always stay on hard-packed surfaces. If you are not sure, get out of the car and check the surface before driving or parking.

I completed an earlier version of this nature guide in 1980. Twenty-seven years have passed since that publication and the valley has experienced dramatic changes resulting from a tripling of the permanent population. Of the ten places that I sent visitors to in 1980, two are only remnants of their former extent. The sand dunes that once occupied the entire valley center are gone except for a tiny section now closed to the public. Most of a huge and diverse stand of cactus known as Devil's Garden has disappeared. Illegal removal by cactus thieves, recurrent fires made worse by exotic weeds and grading for residential development have been the causes of its elimination.

A third site, the Cahuilla Indian Fish Traps, is dying of neglect. The site consists of waist-high stone walls that were built along the shore of an ancient body of water known as Lake Cahuilla. The walls formed U-shaped weirs that aided Cahuilla Indian ancestors in trapping fish swimming along the shoreline. The lake has been gone for hundreds of years but the walls remain. The site is one of the most unique in all of North America. The land on which the traps rest was transferred to the county of Riverside but the county has done nothing to preserve them. As a result of neglect, agricultural grading has removed remnants of the lower traps and unregulated visitation has resulted in the gradual breaking down of the walls. Without county attention, the preservation of these unique features is in serious doubt.

California barrel cacti and brittlebush
in the northwestern Coachella Valley

Although some impressive natural areas have been lost, others have been made more accessible. Most notable is the opening of magnificent Tahquitz Canyon. In 1980 the canyon was closed to the public because unregulated use had resulted in piles of trash and numerous rocks covered with graffiti. Today the Agua Caliente Band of Cahuilla Indians, the owners of the land, have removed the litter and graffiti and opened the canyon to the public via an entry gate and guided tours.

Also of note is the new boardwalk at Big Morongo Canyon Preserve. The wooden trail makes the heart of the cottonwood forest accessible to handicapped visitors. With this and other improvements, the Preserve has become an outstanding user-friendly nature experience.

I hope you enjoy the special places described in this book. They are treasures to experience and most will make indelible impressions. Visiting these areas not only provides one with a better understanding and appreciation of the desert, but the ecological processes that govern all life.

The origin of the name Coachella is obscure. One story claims that it is a bastardization of the Spanish word conchita meaning little shell. Small shells, evidence of ancient Lake Cahuilla, are remarkably abundant in flatland soils of the lower, southeastern end of the Coachella Valley.

An Indian Fish Trap showing the rocks and boulders used to make the original three sides.

How The Valley Came To Be

The Coachella Valley's spectacular scenic diversity results from forces deep beneath the earth's surface. Much of our planet's interior is molten and active. When the molten interior shifts, it can crack and move the thin and brittle crust on which we live. These cracks are referred to as faults and the broken pieces of crust are termed plates. The Coachella Valley straddles two of these plates: the Pacific on the southwest and North American on the northeast. The seam between the plates is named the San Andreas Fault and it nearly divides the Coachella Valley in half.

For the past twenty-five million years, the active molten interior has been sliding the Pacific Plate past the North American at approximately two inches per year. That may not sound like much until one actually feels a plate move during a major earthquake. There have been many memorable quakes that have shaken the Coachella Valley. The largest recent tremor with a local epicenter was the North Palm Springs quake in 1986. It had a magnitude of 6.0 on the Richter Scale. (There has not been a really big earthquake, of 7.0 magnitude or greater, in historical times causing geologists to predict the inevitable "Big One" in the near future.)

The cumulative result of the earthquakes in the region has not only been the northward movement of western California, but also the upthrusting of mountains surrounding the Coachella Valley. These mountains are responsible for the desert conditions that prevail from Palm Springs east to Tucson, Arizona. The high mountains block most westward-moving storms coming in off the Pacific Ocean, creating what geographers call a rain shadow desert.

As if creating the mountains and desert were not enough, earth movements along the San Andreas Fault have also been responsible for several other features including the dropping of the valley floor, the creation of dozens of natural springs (including some hot springs), and the formation of the San Gorgonio Pass that connects the Coachella Valley to the Los Angeles Basin.

The San Gorgonio Pass is a low-lying break in the mountain walls that lie along the western edge of the Coachella Valley. The pass is the accelerator through which prevailing ocean breezes flow into the valley. From time to time, winds can be particularly severe. In June of 1975, the juxtaposition

Desert sunflowers,
Gerea canescens,
on Edom Hill

15

of intense high and low pressure systems resulted in hurricane-like winds that blasted out of the pass at speeds up to 103 miles per hour!

Constant winds emanating from the pass have been responsible for landscape features that, up until very recent times, visually dominated the valley floor. Sand dunes and hummocks were important features over much of the valley for thousands of years. From a distance, their gentle, meandering forms softened the desert landscape and created a welcomed contrast to the hard, angular faces of the enveloping mountains. That scenery, however, is gone today. In the past sixty years, ninety-five percent of the dunes and hummocks that once covered the valley floor have been replaced with residential developments, golf courses, roadways, and a variety of commercial enterprises.

Strong winds at the mouth of the Pass have made it an ideal wind resource area and explains the erection of hundreds of wind-driven generators. The propellor-like blades turn when wind speeds reach 10 miles per hour. On average, each wind turbine produces enough electricity to power forty homes for one year.

Land Below Sea Level

The tectonic forces beneath the Earth's crust have created the San Andreas Fault, uplifted the surrounding mountains, and allowed the eastern valley to subside below sea level.

The eastern surface of the Coachella Valley lies 273 feet below the level of the oceans. This depression is over a hundred miles long and embraces more than 2,000 square miles of below sea level landscape, making it the largest such region in the Western Hemisphere. Geologists have fittingly named the depression the Salton Trough since it was once an enormous, salt-covered playa.

From the outset it must be stated that the Valley is not a valley at all. Real valleys are created by water, by the carving action of rivers over millions of years. The depression known as the Coachella Valley is more correctly referred to as a "rift valley" created by downward movements along

*Sandstorm in the
San Gorgonio Pass*

cracks in the earth's crust. Today, the bottom of the rift valley is occupied by an enormous lake called the Salton Sea.

The Salton Sea is huge, particularly when one considers that it exists in the middle of the Colorado Desert, the driest subdivision of all the North American deserts. One might think evaporation would be so high and rainfall so low that there wouldn't be enough water to form a puddle, much less a sea. Yet from north to south it stretches more than 35 miles and is 15 miles across. There are 360 square miles of surface and 110 miles of shoreline, giving it the largest surface area of any lake in the state.

With a depth that averages around fifty feet, the Salton Sea does not have the largest volume of any lake in California. That honor is bestowed upon Lake Tahoe that has a smaller surface area but is a whopping 1,645 feet deep.

As big as it is, try to imagine an even larger Salton Sea. Imagine one 38 miles wide by 100 miles long with a surface area of over 2,500 square miles—nearly ten times larger that today's Salton Sea. Such an enormous freshwater lake existed in the very same spot as today's sea. The ancient lake, named Lake Cahuilla after the Indian tribe that lived along its shores, was fed by discharge from the Colorado River. The lake would have persisted until today had it not been for sediment buildup at the mouth of the river. The accumulation of sand and mud changed the course of the river and redirected it south into the Gulf of California. By the time the Spanish explorer Juan Bautista de Anza reached the Coachella Valley in 1775, the lake had evaporated away leaving a dry, salt playa in its place.

The ancient shoreline of Lake Cahuilla (not to be confused with the tiny man-made Lake Cahuilla managed by Riverside County in the city of La Quinta) can still be observed today around Travertine Rock. Located adjacent to Highway 86 about a mile south of the Riverside-Imperial County line, Travertine Rock was an island when ancient Lake Cahuilla was in existence. The top of the rock reaches 89 feet above sea level. One of the shorelines of ancient Lake Cahuilla can be seen as a light-dark interface around the rock. (In modern times vandals have partially obscured the shoreline with graffiti.) Another higher shoreline can be seen again on the cliff faces to the west. The rock above the shorelines is light because the waves of the lake eroded away the reddish brown desert varnish. The dark color below the shoreline is travertine, a deposit of calcium carbonate laid down while the rocks were beneath the water. The ancient shoreline of Lake Cahuilla is best seen in the morning while driving south from Coachella along Highway 86.

At 286 feet below sea level, California's Death Valley is lower than the bottom of the Salton Sea. However, the number of square miles of below sea level landscape around the Salton Sea is more than twice that of Death Valley.

The Mountains That Surround

Five mountain ranges enclose the Coachella valley. The lowest range lies to the east and is named the Orocopia Mountains. The Orocopias lie near the town of Mecca. The highest peak in this range is 3,815 feet above sea level and appropriately named Orocopia Peak. To the north lie the Little San Bernardino Mountains. At 5,518 feet, Eureka Peak is the highest point in this range. Most of the this range lies in Joshua Tree National Park. To the northwest rise the San Bernardino Mountains with the tallest point, San Gorgonio Peak, reaching 11,502 feet. This is the highest peak in Southern California.

All three of these mountain ranges (with the possible exception of the Orocopias) comprise what geologists refer to as the Transverse Ranges Geological Province. Unlike most other mountain ranges in North America, the Transverse Ranges have an east-west rather than a north-south alignment. This odd configuration is a result of the San Andreas Fault curving to the west in Southern California. To understand what happens, one must recall that the fault is the meeting place of two great plates of the earth's crust: the Pacific Plate represented on the south side of the valley, and the North American Plate represented on the north side. Normally, the Pacific Plate heads north and slips past the North American Plate. In the Coachella Valley region, however, the fault veers to the west and the two plates collide, crumpling the plate edges into the Transverse Ranges. This process began with the formation of the San

A winter storm arrives in the Coachella Valley

Andreas Fault about twenty-five million years ago and appears to have speeded up in the last five million years.

The southern edge of the Valley is bounded by the Santa Rosa Mountains with the highest point, Toro Peak, rising 8,808 feet above sea level. This range lies within the Santa Rosa and San Jacinto Mountains National Monument and Anza-Borrego Desert State Park. The western end of the Coachella Valley is bounded by the San Jacinto Mountains. Because of their steep slopes and height these mountains dominate the skyline. The northeast face of the range is particularly impressive. In less than seven horizontal miles the peak rises from 800 to 10,804, feet creating the steepest escarpment in North America. No other mountain on this continent rises so high so fast, not even the Sierra Nevada or Grand Tetons.

The San Jacinto Mountains (as well as the Santa Rosa Mountains) are part of what geologists term the Peninsular Ranges Province, one of the largest geological units in North America. Beginning in Mexico at the tip of the Baja Peninsula, the province runs northwest for 900 miles and finally terminates with the San Jacinto Mountains. Only the province's northernmost 120 miles extend into the United States.

The core of the San Jacinto and Santa Rosa Mountains is classified as Mesozoic granitic rock. This is generally a light-colored, once-molten material that can be observed at intermediate to high elevations. The granitic rock resulted from the cooling of magma moving towards the surface about 100 million years ago, during the age of dinosaurs. This granitic intrusion is called a batholith and it formed the core of the Peninsular Ranges that we know today. For tens of millions of years the batholith remained miles beneath the surface.

Over eons of time the rock overlaying the batholith slowly eroded away. The batholith then rose like a ship in water emptying its heavy cargo. Around 30 million years ago enough of the overlying rock had eroded to allow the batholith to rise above sea level and expose its upper portions. When the San Andreas Fault formed approximately 25 million years ago, the Peninsular Ranges Province, including the Santa Rosa and the San Jacinto Mountains, continued to rise due to the collision of the two plates described previously. The crust crumpled, the mountains rose and the valley floor dropped. The process continues today though at a rate imperceptible in a human lifespan.

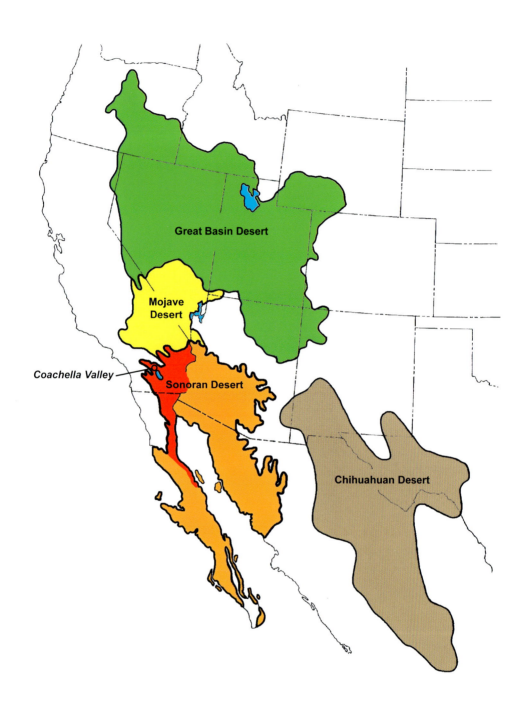

Great Basin Desert

Mojave Desert

Coachella Valley

Sonoran Desert

Chihuahuan Desert

A Desert Climate

The Coachella Valley is located within the North American Desert, a vast area that covers most of the southwestern U.S. and much of northwestern Mexico. Considered in its entirety, the region is distinguished by its spectacular landscapes and impressive plant forms that reveal a diversity found nowhere else in the world. Climatically, the North American Desert is dominated by hot summer temperatures and precipitation that averages less than ten inches per year.

Due to its vastness and diversity, the North American Desert has been divided into four regions: the Great Basin, Mojave, Chihuahuan and Sonoran deserts. It is the Sonoran Desert in which the Coachella Valley is located. From west to east the Sonoran Desert stretches from Palm Springs to Tucson, Arizona. From north to south it begins in Joshua Tree National Park and terminates at the tip of the Baja Peninsula. The region is characterized by low elevations, hot summers and mild winters. Even in January, nighttime temperatures rarely drop below freezing.

Because of differences in topography, climate and vegetation, the Sonoran Desert has also been divided into sub-regions. The California portion of the Sonoran Desert is often referred to as the Colorado Desert, a name derived from the Colorado River which forms its eastern boundary. (The Colorado Desert is shown in red on the map on page 24.) The Colorado Desert is a low-elevation, hyper-arid region that has exceedingly hot summers and generally receives less than five inches of precipitation per year. Temperatures rarely drop below freezing even on winter nights. Unlike the eastern Sonoran Desert of Arizona, the Colorado Desert receives most of its precipitation in winter.

It is the winter climate that attracts residents and tourists to the Coachella Valley. Even in December and January daytime highs typically exceed 70 degrees Fahrenheit. Nighttime temperatures below 40 degrees are unusual and temperatures below freezing are rare. Summer temperatures, however, are hot, sometimes very hot. The record high temperature for the valley was set in the town of Thermal in July, 1995, when a temperature of 126 was recorded three feet off the ground in the shade (the official standard). Maximum daily temperatures in July routinely reach 108 throughout the valley and ground temperatures have been known to approach 200 degrees.

Cloudy days are uncommon in the Coachella Valley. A typical year has more than 325 days of sunshine. It is most likely to be cloudy in winter when storms come in off the Pacific Ocean. There are, however, usually less than twelve such events each year. In times past, storms have been cold enough to bring snow to the valley floor every few decades.

Though the Coachella Valley is just 55 miles long by 18 miles across, there are detectable differences in climate as one moves from west to east or "down valley" in the local vernacular. The western end of the valley is a bit cooler on summer days and a bit warmer on winter nights. The valley's eastern end receives less rainfall and has slightly more sunshine.

Wind varies tremendously depending upon location. Speed and persistence are greatest at the valley's western end, near the San Gorgonio Pass. Winds can accelerate more than four times their initial speed as they funnel through the pass. In May and June, wind speeds in excess of 40 miles per hour are common. Winds decrease markedly as one travels eastward from the pass. Wind speeds also decrease as one moves southward, towards the mountains. Most of the valley cities were initially nestled against the southern mountains to capitalize on the wind protection provided by hillsides and ridge lines. As the human population has grown, however, most of the original "cove communities" have expanded into the windier portions of the central valley.

Climate Data For The Coachella Valley

	Palm Springs	Palm Desert	Indio	Coachella	Mecca
Record High Temperature (in degrees Fahrenheit)	123	123	123	126	126
Record Low Temperature	19	13	13	17	18
Average Daily Maximum Temperature in July	108	108	107	108	109
Average Daily Lowest Temperature in January	43	41	41	37	37
Mean Annual Precipitation (in inches)	5.23	3.15	3.05	3.53	3.33
Elevation (in feet) (* below sea level)	425	240	-10 *	-60 *	-180 *

*Smoke Tree enduring
storm runoff in
Palm Canyon Wash*

Flowers, Cacti and Palms

It seems remarkable that any plant can survive in the arid environment of the Coachella Valley. Yet plants do grow here and in surprising diversity. In fact, there are more than 1,000 plants species that occur on the flats, slopes and mountains of our region. These represent all of the major plant forms including ephemeral wildflowers, grasses, shrubs, trees, ferns and vines.

Most of these plants have the ability to survive long periods of drought through several water-saving adaptations. The cacti, for example, have evolved the ability to accumulate large quantities of water in thick succulent stems. The California barrel cactus (*Ferocactus acanthodes*) is perhaps the most prominent example of this water-holding technique. Its accordion-like stem quickly expands following rains, then slowly contracts as water is used up during periods of prolonged drought. The barrel cactus is popularly believed to provide an emergency source of water to a thirsty traveler. The internal tissues are juicy and there is water to be had. The salt content, however, increases as the cactus dries out and may upset the body's osmotic balance to such a degree that one would be worse off for having ingested the liquid. At best, a traveler might sustain himself for a few hours if the salt content of a particular plant were not excessively high. The barrel cactus is the largest of the more than twenty species of cacti found in and around the Coachella Valley.

The perennial shrubs that dominate the vegetation of the valley have evolved a different mechanism for survival. Creosote bush (*Larrea tridentata*), burrobush (*Ambrosia dumosa*) and brittlebush (*Encelia farinosa*) shed all or part of their foliage when water is scarce. They enter a state of dormancy in which water use is drastically reduced. To the uninitiated, these shrubs may appear dead. Not until winter rains have fallen do they return to "life" and produce new leaves.

A third strategy for survival has been evolved by the ephemeral wildflowers that usually live for only a few months or even weeks. These plants survive the hot, dry seasons as seeds until the onset of winter rains. In late winter and spring, should ample precipitation fall, the valley floor becomes carpeted with the beautiful flowers of sand verbena (*Abronia villosa*), dune primrose (*Oenothera deltoides*), and desert sunflower (*Gerea canescens*).

Beaver-tail Cactus,
Opuntia basilaris, *in bloom*

*The Coachella Valley has a wild, native orchid. True to its name, the stream orchid (*Epipactis gigantea*) is restricted to the vicinity of streams and seeps where the orchid has access to a permanent water supply. Individual plants can reach three feet in height.*

A fourth category of plants are those restricted to dry washes. Here surface water flows only after intense storms that send runoff down the canyons and out onto alluvial fans. Water persists for several days or even weeks in the soil beneath the wash beds. In these environments several tree species can be found—to visitors they may seem more like large shrubs—including smoke tree (*Psorothamnus spinosus*), palo verde (*Cercidium floridum*), ironwood (*Olneya tesota*) and desert willow (*Chilopsis linearis*).

A final group of plants are those restricted in their distribution to permanent springs and streams where perpetually damp soils allow the growth of plants that show no particular water-conserving adaptations. The presence of the San Andreas Fault and numerous associated splinter faults and fractures have resulted in an unusually large number of such environments in the Coachella Valley. Foremost on this list of oasis plants is the desert fan palm (*Washingtonia filifera*). This is North America's largest palm and reaches its greatest abundance in the Coachella Valley. It is also a species almost entirely restricted to the Colorado Desert. A number of other tree species are found in the valley's oasis environments including true willow (*Salix* spp.), Fremont's cottonwood (*Populus fremontii*), California sycamore (*Plantanus racemosa*), and white alder (*Alnus rhombifolia*). Mention must also be made of the honey-pod mesquite (*Prosopis glandulosa*), an often enormous shrub that is encountered in many oasis environments.

A wide variety of desert plants are edible. The list includes cactus fruits from the beaver-tail, prickly-pear and barrel cacti; flowers and fruits of various yuccas; and blossoms and seed pods of the mesquite. Conversely, some of our desert plants, such as locoweed and jimsonweed, are toxic.

Dune Primrose, Oenothera deltoides, *common in areas of windblown sand*

Valley Wildlife

The tremendous variety of habitat types in and around the Coachella Valley support equally diverse animal life. Four hundred and fifty vertebrate animals have been recorded from the region including 8 kinds of amphibians, 40 reptile species, 347 species of birds, and 58 species of mammals. Included within this list are one reptile and one amphibian species found nowhere else in the world.

Without question the Coachella Valley's most picturesque animal is the bighorn sheep, *Ovis canadensis*. A mature ram may reach 200 pounds and, with its muscular torso and massive curled horns, is the largest native animal. Bighorn are partial to the lower slopes of mountains and are found in all the surrounding ranges. They are most likely to be seen while hiking in the Santa Rosa Mountains on the south side of the valley.

The cat-sized kit fox is occasionally observed scampering across roadways at night, particularly on the north side of the valley. These large-eared canines become active at dusk as they search for kangaroo rats and rabbits, their chief prey. Ringtails, bobcats, gray foxes and mountain lions are among the larger predators within the Coachella Valley region, but these animals are rarely seen. The ubiquitous coyote is the carnivore most regularly observed by residents and visitors. It can be found everywhere in the region and most likely encountered at dawn and dusk.

Exciting as it is to see a bighorn or coyote, it is the smaller animals that are most abundant and most often encountered. The antelope ground squirrel (*Ammospermophilus leucurus*), distinguished by striped sides and a tail that is conspicuously white underneath, is common in areas of coarse soil and rock away from urban areas. Jet black ravens (*Corvus corax*) are observed on most any day, usually gliding overhead in search of food scraps around homes and businesses or perhaps a jackrabbit carcass on the road. Roadrunners (*Geococcyx californianus*) nest in trees and shrubs surrounding homes on the outskirts of residential areas where insects and side-blotched lizards (*Uta stansburiana*) are common. The latter reptile is characterized by a two-inch body adorned with light stripes (female) or speckled with blue and yellow (male). Individuals can be seen basking on rocks on any calm day, even in winter when all other reptiles are in hibernation.

Female bighorn sheep and lamb,
near Palms to Pines Highway,
Santa Rosa Mountains

Many visitors to our area fear they must contend with venomous creatures such as rattlesnakes, scorpions, or centipedes, but in truth these animals are rarely encountered. The vast majority of desert residents never see any of the five species of rattlesnake known to occur within the region. Scorpions are nocturnal and active when people are at home asleep. (None of the scorpion species within the Coachella Valley are considered dangerous to humans.) In fact, the most unusual creature that visitors might encounter is a large arachnid known as the tarantula. Males are often seen in fall and early winter while searching for mates. Ferocious as they may appear, tarantulas are quite harmless.

In large part, the diversity of animal life that exists within the Coachella Valley is a testament to the superb behavioral adaptations these animals have made in responding to their environment. Unlike plants that have no choice but to endure environmental extremes, animals are mobile and can move into more favorable micro-environments when advantageous. Many desert animals are nocturnal, becoming active at night when temperatures are lower and the air more humid. Small animals such as pocket mice, kangaroo rats and ground squirrels (and even larger animals, such as bobcats) burrow underground where temperatures never rise above 80 F. White-crowned sparrows (*Zonotrichia leucophrys*), western bluebirds (*Sialia mexicana*) and ruby-crowned kinglets (*Regulus calendula*) spend only the winters in the Coachella Valley. With the onset of warm weather they fly away to milder climates.

*Four officially endangered or threatened animal species are found in the Coachella Valley: the desert pupfish (*Cyprinodon macularius*), desert tortoise (*Gopherus agassizi*), desert slender salamander (*Batrachoseps aridus*) and Coachella Valley fringe-toed lizard (*Uma inornata*). The latter two animals are found nowhere else in the world. Several other subspecies or populations within the Coachella Valley region have also been declared threatened or endangered by the U.S. Fish & Wildlife Service. In the Coachella Valley, loss of habitat due to urban sprawl is the principle threat to the survival of these animals.*

Male Costa's Hummingbird,
Calypte costae

The First People

Celebrities, tourists and settlers were not the first humans to occupy what is today Palm Springs. Fourteen thousand years ago, during the Pleistocene ice age, Asian people crossed a natural land bridge made of snow and ice that connected Siberia to Alaska. Their descendents would eventually migrate throughout North and South America and occupy the Coachella Valley not less than three thousand years ago.

The Cahuilla Indians (pronounced Ka-wee'-yah) are the descendents of these first North Americans and were the people inhabiting the valley when Spanish explorers arrived. Language patterns indicate that the Cahuilla are most closely related to the Aztecs of Mexico and the Hopi and Pima of Arizona. Cahuilla language and cultural traditions were practiced by the people who lived in western Riverside County and southern San Bernardino County, California, an area of about 3,500 square miles.

Since their first encounter with Europeans in 1775, Cahuilla traditions have become increasingly overshadowed first by those of the Spanish and subsequently by the "Americans." It has been estimated that prior to contact with persons of European ancestry, the Cahuilla numbered between 4,000 and 7,000 persons. Today, there are approximately 800 individuals who claim Cahuilla ancestry.

The Coachella Valley city of Indian Wells is named after water-filled pits created by Cahuilla Indians. Prior to contact with Europeans in the late 18th century, walk-in wells were dug with wooden shovels. The wells reached ground water that rose to within 10 feet of the surface.

A large body of evidence indicates that the Cahuilla were a people of impressive resourcefulness and industry. The Desert Cahuilla (a popular name given to Cahuilla living in the Coachella Valley) were one of the few North American tribes to dig wells. These excavations were often large enough for villagers to walk down to the water level, sometimes eighteen or more feet below the surface. They were also one of the few groups in North America to utilize permanent structures to capture fish, fish that were plentiful in a lake that once covered much of the lower valley.

Dolores Patencio, a Cahuilla woman, grinding seeds in a hopper. Photograph taken around 1930.

The Cahuilla were occasionally known to have built reservoirs filled with water from wells or streams. The discovery of what may have been pre-historic canals suggests that the water was used to irrigate crops. If the Cahuilla practiced irrigated agriculture, this fact would add to the increasing body of evidence that the Cahuilla have a rich heritage of cultural and technological accomplishments.

Other notable features of the Cahuilla Indians are their extensive oral traditions, the excellent basketry of the Cahuilla women and perhaps most importantly, their success in adapting to the rigorous extremes of the desert environment.

Palm Springs received its name from hot springs beneath the site of the present-day Spa Hotel in downtown Palm Springs. Cahuilla Indians were first to discover and use the therapeutic waters long before Europeans arrived.

By all rights, the Cahuilla should not have been able to colonize the desert lands of the Coachella Valley. The valley is among the hottest and most arid environments in North America with seemingly little food and no obvious sources of water. The region is, however, a remarkably diverse environment that on closer scrutiny provides abundant water via mountain streams and numerous springs along the San Andreas Fault.

Diverse habitats provided an unusual array of plant and animal resources that were fully exploited by the Cahuilla. Although Cahuilla villages were stationary, groups of individuals would travel up to fifteen miles on food-collecting trips. In springtime, cactus buds, blossoms and fruit were gathered on alluvial fans surrounding the valley floor. In summer, mesquite pods were harvested from plants growing along the San Andreas Fault as well at canyon springs. In fall, family groups hiked high into the mountains to collect acorns and nuts of the pinyon pines. Animal resources were also diverse from quail and jackrabbits on the desert floor to deer and bighorn in the high country. Families acquired exotic foods through trade with other villages. Extensive trade networks connected the Cahuilla with tribes in Arizona, Nevada, Mexico and coastal Southern California.

Broken remains of ancient Indian pottery laying exposed on the desert surface

Palm Canyon

What you will experience: *largest palm oasis, cathedral-like groves, running water, bedrock mortars*

It is midday but there is no sunlight. High above, a canopy of leaves separates me from the sun. There are many birds but the roar of rushing water drowns out their songs. Everywhere it is green, verdant, alive. If I close my eyes it is easy to imagine I am in the deep, moist forests of the Pacific Northwest or even the Amazon. In fact, I am in the Colorado Desert of southeastern California, second only to Death Valley as the hottest and driest place in North America. How can this be?

Hidden away in the canyons, ravines and badlands of arid California are more than a hundred wild palm groves, islands of green in an otherwise stark and foreboding landscape. Water is the key to the existence of these oases. In some places it emerges from faults—cracks in the earth's crust that allow groundwater to percolate to the surface. Elsewhere, runoff from lofty mountains creates streams that flow into the desert. In both cases it is a single species of palm that turns the desert into jungle.

The palm in question is the desert fan palm, known to aficionados as *Washingtonia filifera* and named by a European to honor America's first president. It is no ordinary palm. With it comes a string of unique characteristics. For example, it is the only native palm in the western United States and is the largest palm in North America. Research has also shown that it is the most cold tolerant palm in the world. It has a distinctive appearance resulting from dead fronds that remain attached to the trunk throughout its hundred-year lifespan. None of the other 2,800 palm species possess this trait.

Though most palm oases are small with less than twenty trees, a few are large, harboring a hundred or more stately individuals. There is just one oasis where palms number in the thousands and it is aptly named Palm Canyon. For twelve miles, this fault-created ravine separates the Santa Rosa from the San Jacinto mountains. On the trail above the canyon, hikers look down upon a ribbon of green created not only by palms, but by cottonwoods, willows, and sycamores as well. On the trail it is hot and dry, but in the canyon bottom there is abundant shade and running water

*Desert fan palms
in Palm Canyon*

In all of the other desert regions on the planet, there is no natural palm oasis that is so vast with so many palms.

Though I had visited Palm Canyon many times to view the deep, crystal clear pools and occasional waterfalls, I returned in 1985 on a serious mission. Two early researchers of palm oasis ecology, Richard Vogl and Lawrence McHargue, had written papers suggesting that desert fan palms were in trouble, that they might be heading towards extinction. Off-road-vehicles, land development, and air pollution were cited as possible threats. Their fears, however, were not documented and a formal research project was necessary to determine the health and vitality of palm oases.

Research costs money. Lucky for me Steve McCormick and the Nature Conservancy became involved with the project in its intial phases. Through their efforts, I received a grant from the Richard King Mellon Foundation of Pittsburgh, Pennsylvania. The grant would allow for field investigations that would help the scientific community better understand the ecology of palm oases and determine the status of palm populations.

After three years of field investigations, it turned out that Palm Canyon Oasis, like most other palm oases that our research team visited, appeared to be getting bigger. In the 1950s, the late Randall Henderson estimated that the number of adult palms in the canyon approached two thousand. But when naturalist Dave Mathews and I hiked the twelve miles down the canyon in 1984, we counted 2,511 palms, an estimated 20% increase. Even more palms grace Palm Canyon today.

One explanation for the increase in palm numbers was fire. Once every few decades, lightning can be expected to strike a tall palm tree. The fronds are highly flammable. The late Chris Moser, of the Riverside Municipal Museum, actually observed a lightning bolt strike a *Washingtonia* palm. He related how the top of the palm literally exploded when struck, sending flaming leaves in all directions. Such an event could be expected to set an entire palm grove on fire. Most oasis fires today are not started by lightning, but by humans—either accidentally or as acts of vandalism. Such was the case with in the Dry Falls Fire of 1980. Two small boys, playing with matches, started what became an enormous fire that eventually charred some 60,000 acres. The fire raced through much of Palm Canyon, transforming lush trees into blackened poles.

Fortunately for today's visitors to Palm Canyon, the desert fan palm has a remarkable ability to survive a conflagration. When a palm oasis burns,

*Main grove of
desert fan palms
in Palm Canyon*

adult palms typically lose their leaf skirts, trunks are charred, and green crowns destroyed. Within months, however, the trees reveal that they are very much alive as they produce an entire new crown of green fronds. Nearly all of the adult palms in Palm Canyon came to life after the fire. The ability of a palm to survive burning is related to its family history. Many, if not most members of the Palm Family are adapted to wildfires. Palms are monocots, which means their sensitive vascular tissue is strewn throughout the moist trunk, not confined to a narrow cambium layer beneath the bark, as is the case with pine and oak trees. Many fires are sufficiently hot to destroy the vascular tissue of these latter trees but rarely is a fire hot enough to kill the tissues deep in the center of a palm.

Not only does *Washingtonia* survive fire, it thrives in a fire regime. Hidden Palms, on the floor of the Coachella Valley some fifteen miles from Palm Canyon, burned three times between 1939 and 1979, the greatest fire frequency endured by any oasis. Hidden Palms responded with the highest ratio of young to mature palms of all the oases I studied. By 1983, there were 264 young palms compared with 191 adults, a ratio of seven to five. Fire promotes the establishment and growth of young palms by removing competing species. Could the burning of Palm Canyon in 1980 account for the impressive increase in palms?

The answer was no. The charred trunks that hikers encounter in Palm Canyon today were the same adult palms that were there before the fire. Even with their initial growth rate of a foot each year, the palms could not have reached their height of thirty, forty or even fifty feet in the four years that elapsed before we arrived to census the palms. We are still working on the explanation for the increase in palm numbers, not just in Palm Canyon, but throughout the Colorado Desert. There are hints that global warming may be a factor.

Not surprisingly, the combintation of palms, water and shade attract not only people but a variety of wildlife species. Over the course of a year, ornithologist Theo Glenn and I counted eighty species of birds including the beautiful orange and black hooded oriole, the endangered Bell's vireo, and the strikingly blue lazuli bunting. Near the stream, we found the cryptically colored California treefrog that is almost impossible to see when it rests upon a granite boulder. Early morning visitors may encounter a harmless kingsnake crawling across the oasis floor, a wading great blue heron, or a bighorn sheep that has come to drink. Wildlife is abundant in Palm Canyon, attracted by the wild foods, water and the shelter provided by the palms.

Three thousand years ago Cahuilla Indians found the canyon every bit as appealing as modern-day tourists and local residents. They took up residence in and around the canyon, established villages and began tending the palms to increase fruit production. Though today the Cahuilla are perhaps best known as operators of the Agua Caliente and Spa casinos, in ancient times their ancestors took advantage of all the natural products offered by the oasis: a reliable source of water, palm fronds for building rainproof dwellings, fruits and seeds from both the palms and mesquite, plus a variety of game. Today, evidence of Cahuilla presence can be found throughout Palm Canyon in the form of bedrock mortars situated on large, flat rock outcrops. The mortars resulted from the grinding of both palm and mesquite fruits with a large, cylindrical stone called a pestle. Over decades, even centuries, a hole was worn in the bedrock, in some cases up to ten inches in depth. Hundreds of these exist in Palm Canyon and many can be found near the main trail.

How to get there

This spectacular environment is located just six miles south of downtown Palm Springs. The entrance gate to Palm and its associated canyons is located three and one-half miles south of Tahquitz Canyon Way and opens every day at 8:00 a.m. It can be accessed via South Palm Canyon Drive. Palm Canyon is part of the Indian Canyons Tribal Park of the Agua Caliente Band of Cahuilla Indians. A nominal fee is charged for each adult admission.

Bedrock mortar used for grinding plant fruits and seeds

Palm Springs Aerial Tramway

What you will experience: *spectacular view of the Coachella Valley and California deserts, dense coniferous forests, mountain animals, much cooler temperatures*

One of the wondrous things about living in Southern California is the remarkable landscape diversity. In just over two hours a traveler can drive from the Pacific Ocean shoreline up to snow-covered mountains then down through oak-studded grasslands into the Sonoran Desert. No other region in America offers so much, with so little effort.

In my mind the best way to experience this diversity is to ride from desert floor to mountaintop in just ten minutes aboard the Palm Springs Aerial Tramway. Since 1963, the tramcars have been hauling passengers nearly 6,000 feet skyward, from the cactus-covered desert floor to pine and fir forests in the San Jacinto Mountains. The ride is amazing but all too brief. Fortunately, everyone gets two rides, one up and one down.

The tram adventure begins not in a tramcar but in one's own automobile— as you leave Highway 111 and turn west onto Tramway Road. This intersection is situated at approximately 750 feet above sea level and lies at the edge of the great Sonoran Desert—the same desert that extends eastward all the way to Tucson, Arizona.

No, you won't see saguaro cactus in the local version of the Sonoran Desert, but you will see many other desert plants including golden cholla, beavertail cactus, and the desert's most important perennial, the creosote bush. All these plants are growing on the lower reaches of the boulder-strewn slope that lies at the mouth of Chino Canyon, the host canyon of the tramway. Geologists refer to the slope as an alluvial fan and there's no finer example of such a feature in all of the Coachella Valley. It is steeper and covered with more and larger boulders than any other alluvial fan in the region.

The fan was created entirely by flowing water in a process that began more than five million years ago when Chino Canyon was created. Runoff from storms has carved and carried everything from tiny grains of sand to house-sized boulders down the canyon. Though most downward

Coniferous forest on the slopes of San Jacinto Peak

49

transport of rock has been slow, torrential downpours and floodwaters have, at times, moved car-sized boulders more than a mile down the fan. Though such large events are infrequent, given the millions of years of the canyon's existence, ample time has elapsed to form the impressive Chino Canyon Alluvial Fan.

The steepness of the fan makes driving up Tramway Road a little like the methodical, beginning climb of a roller coaster. For each mile ascended there is about a 500-foot gain in elevation, two-degree drop in temperature, and much evidence of increasing precipitation. The evidence is in the form of changing plant life. New plants appear—species like cat's claw acacia, the yucca-like Parry nolina, and, on south-facing slopes, the barrel cactus. Occasional wildfires ravage the area and burned slopes can usually be seen on the south side of the canyon. Fires have become more frequent in recent years as persistent exotic grasses and wild mustard plants now provide a continuous corridor of fuel from the mountains to the edge of the desert.

At about mile three Tramway Road crosses a near-perennial stream, an obviously rare phenomenon in the desert. The vegetation becomes lush as the abundant moisture allows the growth of thick stands of sycamore, willow and cottonwood trees. Half a mile further the stream continues to flow but usually does so underground and spreads from wall to canyon wall. A sea of green sweeps across the canyon at this juncture—a sea comprised of not only the trees mentioned earlier but also wild grapevines and desert fan palms. This is the Cienega Oasis, home to the largest concentration of Least Bell's Vireo anywhere in the California deserts. This vireo is an endangered species and locals hope the oasis will, someday, be designated a nature preserve. That decision, however, rests with the wildlife regulatory agencies and the landowners. Help may be on the way as the oasis has been included within the Coachella Valley Multiple Species Habitat Conservation Plan.

Driving ever upward, the canyon narrows and this is both the last and best chance of seeing the magnificent bighorn sheep prior to boarding the tramcars. Rams (males) can approach two hundred pounds in weight and can be identified by their curled horns. With declining numbers due to disease and habitat loss, it is the lucky visitor that glimpses one of these mammals.

Once on board the Palm Springs Aerial Tramway the vistas become spectacular. The tramcars slowly rotate and so no matter where you

Palm Springs Aerial Tramway
car heading up Chino Canyon

stand you'll eventually be able to look in every direction. On most any day the Santa Rosa Mountains and Toro Peak can easily be seen to the south. The Salton Sea is located to the east. The Little San Bernardino Mountains and Joshua Tree National Park are easily viewed to the north. West of the Park stands San Gorgonio Peak. At 11,502 feet it is the tallest mountain in Southern California.

At the top, walking out of the tram car and into the wilderness of San Jacinto State Park rapidly demonstrates the impact altitude has on the natural world. Gone are cacti and creosote bushes and in their place are pine and fir trees. This almost unbelievable change has happened not by driving a thousand miles north to Canada but by riding two and one-half miles in a tramcar—a tramcar that rises nearly six thousand feet in just ten minutes!

At the Mountain Station temperatures are generally thirty degrees cooler than in Palm Springs. Precipitation is also much greater, usually more than four times the amount that falls on the desert below. This cooler, wetter climate enables coniferous forests to dominate the landscape. In winter, snow blankets the ground and air temperatures can be expected to regularly drop below freezing at night. By June, most of the snow has melted allowing the emergence of wildflowers. Many species continue to bloom into August.

If wildlife is your thing then you'll be pleased at the top. In place of cactus wrens and roadrunners, expect to see Steller's jays and Clark's nutcrackers. Discerning birdwatchers may also glimpse red-breasted and pygmy nuthatches, hairy woodpeckers, and both western and mountain bluebirds. Reptile fanciers can expect sagebrush lizards, western fence swifts and possibly a mountain kingsnake, easily the most beautiful reptile in the San Jacinto Mountains. If you think you have left rattlesnakes behind as you rode up the tramway, think again. The western rattlesnake has been found as high as 9,000 feet in the San Jacinto Mountains.

Although 99% of the plant species change from desert to mountaintop, the change in animal life is not quite so dramatic. Plants can't hide from inclement weather but animals can. The shelter sought by Beechey ground squirrels and desert woodrats protects them from either excessive heat or cold and so they are found from desert floor to the tram's Mountain Station. Other animals are large and mobile and can move up and down the mountain depending upon the weather. Bighorn sheep typically occupy the lower, desert slopes behind Palm Springs but have been known to

Snow covers the ground from late fall into spring at the top of the Tram

Mule deer doe

climb as high as 8,500 feet during the summer months, presumably in search of better food. Mule deer are also known to migrate up and down mountains with the seasons.

The more energetic visitors to the mountaintop may want to take the trail to San Jacinto Peak, a round-trip distance of about eleven miles. The mountain station is situated at 8,516 feet and the peak reaches 10,831 feet, an elevation gain of 2,315 feet. Be prepared for a hike of seven hours. Along the way you will pass by all of the high country conifers. These include the white fir (*Abies concolor*) with its short solitary needles, Jeffrey pine (*Pinus jeffreyi*) with its needles bundled in threes, lodgepole pine (*Pinus contorta*) with two short leaves to a bundle, sugar pine (*Pinus lambertiana*) identified by its bundles of five needles all three inches in length and limber pine (*Pinus flexilis*) with five, two-inch leaves to a bundle. The limber pine is found all the way to the top of the peak and so contrary to what is often said, there is no treeless, alpine zone in the San Jacinto Mountains.

Visitors to the Coachella Valley routinely enjoy the experience of traveling from desert floor to mountaintop. It is one of those must-do activities while in the valley. I continue to be surprised, however, by the number of valley residents who have never ridden the Palm Springs Aerial Tramway or who have only experienced it once in their lifetime. I must confess, as a naturalist, Palm Springs seems unimaginable without the tram. Far and away, it provides the greatest environmental diversion possible, and with so little effort.

How to get there

The Palm Springs Aerial Tramway is located in Palm Springs. From downtown, drive north to Tramway Road. Turn left and drive approximately 4 miles to the Tram's Valley Station. Visitors coming from Los Angeles via Interstate 10, exit on Highway 111 to Palm Springs. Drive 8 miles to the second traffic signal and turn right on Tramway Road. The Tram's Valley Station is 4 miles up Tramway Road.

Tahquitz Canyon

What you will experience: *Valley's most impressive waterfall, dramatic rock formations, ranger tours, sighting of a dipper— the bird that walks underwater*

Picture a perpetual waterfall as tall as a two-story building. Clear water plummets into a pool so deep waders become swimmers if they dare approach the cascade. So much water plummets over the cliff that daredevil bathers are pushed to the bottom. Spray from the falls makes shoreline spectators cool and damp, even on the hottest of days.

Is this Hawaii? No. Somewhere in a South American rainforest? Try again. A canyon along the east face of the Sierra Nevada? Closer. The place is Tahquitz Canyon, not even a mile from downtown Palm Springs. It's hard to believe that such a place exists, nestled as it is into cactus-covered hillsides and so readily accessible. Spectacular waterfalls should only be reached by long, arduous hikes and should exist anywhere but in a desert. But this is the Coachella Valley, a region as rich in landscape diversity as any place in North America. The existence of a place like Tahquitz Canyon (pronounced Tahkeets) should therefore come as no surprise.

I first entered Tahquitz Canyon thirty-five years ago. Back in 1975, the canyon was closed to hikers though few heeded trespassing signs. Unregulated use resulted in piles of trash and graffiti on boulders and rock walls surrounding the waterfall. A few scary looking souls lived in the canyon complete with cardboard campsites. In short, the area was seriously neglected.

Today, that has all changed. Tahquitz Canyon has been restored to its original granduer by the Agua Caliente Band of Cahuilla Indians. The Tribe always owned the canyon—it was part of their reservation land. Until quite recently, however, they did not have the means to regulate its use. But between federal government grants and proceeds from tribal business enterprises management of the canyon became possible and the canyon was opened to the public.

Hikers stepping across stream in Tahquitz Canyon, falls in background

57

Today, the refuse is gone, the graffiti has vanished and no scary vagrant will accost you. In fact, I can say without hesitation that the Tahquitz Canyon experience should be the first place you visit when arriving in the Coachella Valley or entertaining out of town guests.

My latest sojourn to this area became memorable the very moment I turned off the western terminus of Mesquite Avenue and into the Tahquitz Canyon Tribal Park. Yellow-flowered brittlebushes were everywhere, turning the normally drab landscape into a celebration of nature. The paved road wound through naturally varnished boulders and into a parking lot sufficient in size to handle any holiday crowd. From there I walked up to the Visitor Center and paid the entrance fee. The fee includes viewing a short film on the legend of Tahquitz (I won't spoil the experience by telling you why people have vanished in the Canyon), and the services of a Tribal Ranger guide whose knowledge is unsurpassed. I have gone on many guided natural history tours in my life but I can't recall ever having been with someone so informative.

Once you have met your ranger guide, and received a brief introduction to the two-mile hike, you're onto the trail heading to Tahquitz Canyon Falls. At appropriate breath-catching intervals, your ranger will stop and discuss the plants and animals encountered—and there are plenty to talk about. The mouth of Tahquitz Canyon is one of the last places in the Coachella Valley where one can find the chuparosa shrub still in abundance. Red-flowered plants are uncommon in the desert and *Justicia californica* (its scientific name) is a visual respite from the dominant yellow-flowered plants.

Cacti are also common at the canyon mouth. Most notable is the barrel particularly in the spring when its yellow blossoms form a circular crown on the top of the stem. (This insures that pollinating insects will see the flowers from any direction.) By weight, the barrel cactus is the largest cactus species in the region. Historically it has been known as a source of life-saving water for lost travelers. Research has shown, however, that the salt content of the water in the cactus is too high to be of any benefit to humans. Other cactus species found near the trails include the flat-padded beaver-tail cactus, the multi-branched golden cholla and the beautifully flowered calico cactus.

Of course it is the water that attracts humans to the canyon. Tahquitz Creek runs year-round in years of abundant winter precipitation and

Brittlebush, Encelia farinosa, *in bloom in Tahquitz Canyon*

rarely dries up completely, even at summer's end. The reliable water source enables the existence of both Fremont cottonwoods and western sycamores, tree species that require permanently moist soil. At the falls exist two monster sycamores with trunks more than four feet in diameter. These trees have seen better years but are rich in history. Years ago they supported long ropes that were used by visitors to swing, Tarzan-style, from the cliff face out over the pond. The fun part was not the swing but the plunge into the deep pool after releasing the rope. Today the ropes are gone but the broken limb scars bear testimony to decades of use by each Tarzan impersonator.

The falls are impressive by even Hawaiian standards. I estimated that about 10,000 gallons per minute were plummeting down the sixty foot cliff during my most recent visit. The roar of the crashing water prevented talking and the mist lowered the temperature by fifteen degrees according to a pocket thermometer. For the brief period I spent watching the falling water, it certainly didn't feel like I was in a desert environment.

Animal life in the canyon diverges radically from the fauna of the arid hillsides. Through most of the late winter, spring and early summer expect to hear the quacking of California treefrogs (*Hyla cadaverina*) or, if one is observant, actually see one clinging to a boulder near the water's edge. These two-inch amphibians have enlarged pads at the end of each toe that enable them to cling to the slickest of surfaces, even glass. Aquatic garter snakes (*Thamnophis atratus*) patrol the shallows hoping to make a meal of either the treefrogs or their tadpole larvae.

It is the lucky visitor that gets to see the American dipper (*Cinclus mexicanus*), a chunky, dark gray bird that often nests under waterfalls. The sharp claws and powerful feet enable it to walk on the slickest rocks and even on the bottom of a stream. They stay busy under the surface looking for small aquatic animals including the larvae of insects. In the Coachella Valley region dippers occur only in a few canyons, including Tahquitz, where there is a permanent supply of clear, flowing water. These birds are far more common along streams draining the Sierra Nevada.

One of the nice aspects of the guided walks is the absence of backtracking. The walk in is different than the walk out, though the distance is about the same. With the guide, one hears how Native Americans relied upon the creosote bush as a cure or to relieve the symptoms of several ailments. We also learned the importance of mesquite as a food plant, particularly during dry years when many other desert plants failed to produce fruit

Western sycamores,
Plantanus racemosa,
in Tahquitz Canyon

and seeds. On the way out I took advantage of a trial program that allows visitors to walk in and/or out by themselves, without a guide. (This may become a permanent offering.)

Perhaps it was because I was alone that I chanced upon a speckled rattlesnake crossing the trail in front of me. Startled, it rattled the entire time it took to maneuver under a bush. It coiled, rattled more, and dared me to get closer. Speckled rattlesnakes are loath to strike but have one of the more potent venoms among the dozens of rattlesnake species.

As I walked out of the canyon I turned back to view, for one last time, the magnificent mountainsides and jagged rock outcrops of Tahquitz Canyon. The scant vegetation revealed the geological forces that have operated here for millions of years. It also confirmed that I was in a desert place, regardless of the abundance of water flowing through the canyon.

How to get there

In Palm Springs take highway 111 (South Palm Canyon Drive) to Mesquite Avenue. Turn towards the mountain (west) and drive straight to the entrance gate. The gate opens at 7:30 a.m. and closes at 5:00 p.m. Call the Agua Caliente Band of Cahuilla Indians (760-416-7044) for more information or to reserve a hiking time.

Palms To Pines Highway

What you will experience: *Scenic vistas of valley floor, rapidly changing environments, desert agaves in bloom in late spring, cooler temperatures*

Less than a week after finishing my first year of college I jumped into my old, generally reliable VW Bug and headed across the United States. At nineteen, I had never been further east than the Colorado River and yearned for new landscapes. I was not disappointed. Heading towards the Atlantic Ocean, I first marveled at the strange saguaro cacti of Arizona, was then awestruck by the enormity of New Mexico's Carlsbad Caverns, and surprised that snow still covered Rocky Mountain peaks in June. As the song had said, America really was "the beautiful."

Then something went very wrong. The mountains disappeared. Everything became flat. There wasn't a single natural feature to view. The only thing to view was an occasional billboard, grain silo and power pole. What happened was that I drove into Kansas, a scenically challenged place, to say the least. I survived the day-long drive through the state vowing to never again endure such monotony.

My problem with Kansas is that I'm from California. On a clear day, there is no place in my state where one can stand without seeing mountains or some other impressive landscape feature. Visually speaking, I'm very, very spoiled. In the Coachella Valley we are surrounded by mountains and there is always different scenery, regardless of the direction one looks.

To better illustrate this point, there is a driving trip that grandly displays our scenic diversity, one that is the antithesis of driving through Kansas. By traveling up Highway 74, locally known as the Palms To Pines Highway, one can view ten times more visual drama in an hour than in an entire day's driving through the Jayhawker state.

If you're contemplating an excursion up Highway 74, set aside at least one-half day. Start in the city of Palm Desert, at the toe of the Santa Rosa Mountains where the elevation is about 1,000 feet. This juncture is

Highway 74, the Palms to Pines Highway, winds up the Santa Rosa Mountains.

one of the northern boundaries of the new Santa Rosa and San Jacinto Mountains National Monument and the location of the Monument's Visitor Center. Exhibitions at the center provide an orientation to the natural and human history of the region. Outside the building are dozens of native plant species including blue palo verde (*Cercidium floridum*), smoke tree (*Psorothamnus spinosus*) and chuparosa (*Justicia californica*). Many of these are identified with labels and each will be encountered during the drive up Highway 74 and into the heart of the Monument.

The second stop is just across the highway from the visitor center, a place where there are two canyons for hiking. Dead Indian Canyon heads off to the west and Carrizo Canyon starts off to the south. Both canyons lead to springs that support stands of the native desert fan palm, *Washingtonia filifera*. These, and several other palm oases in the immediate region, are responsible for the palm half of the highway's name.

To protect dangerously low numbers of endangered Peninsular Bighorn Sheep, trail access in the Monument is restricted to certain months. Check with visitor center personnel or the Bureau of Land Management (BLM) in Palm Springs to find out when trails are open for hiking.

In the first mile of driving up the Highway's grade you will discover rocky hillsides dominated by creosote bush (*Larrea tridentata*) and brittlebush (*Encelia farinosa*), shrubs indicative of the Sonoran Desert. In a spring following a wet winter these hillsides explode with yellow flowers produced by the two shrubs. Observant travelers will also spot the ocotillo (*Fouquieria splendens*), an odd-looking plant that produces hoards of tall stems but almost no branches. The tip of each stem is aflame with red flowers beginning in March—regardless of the amount of winter rain.

At 2,400 feet above the level of the oceans, and four driving miles south of the monument visitor center, the first official viewpoint is encountered. To enter the parking lot you will be turning left in front of oncoming traffic (be very careful). The view from Vista Point is breathtaking. From the viewpoint, one can see the entire desert floor and most of the cities. You can also spot Southern California's highest mountain—San Gorgonio Peak(to the northwest), Joshua Tree National Park's Little San

Mojave Yucca,
Yucca schidigera

Bernardino Mountains (straight north), and the infamous San Andreas Fault (running from left to right through the valley center) are all easily seen from the viewpoint. Continuing our ascent, we turn left out of the parking area (with care) and head toward the Cahuilla Tewanet Overlook. Along the way another interesting desert plant becomes conspicuous on either side of the road—the desert agave, *Agave deserti*. In proportion and leaf arrangement the agave resembles a giant artichoke and, like the artichoke, it is edible when properly cooked.

The agave was one of the most important plant groups for Indians living in the desert Southwest. Not only did Native Americans use agaves for food but as fiber to make threads and rope. The yucca (the short-trunked plant with bayonet-like leaves seen along the highway) also provided both food and fiber. In May, watch for the agave's tall, yellow-flowered stalk that can extend ten feet above the leaf cluster.

At 3,800 feet we arrive at the Cahuilla Tewanet Overlook. The United States Forest Service has created a short nature trail here with informative labels that describe the original occupants of the area, the Cahuilla Indians. Plant life along the trail is typical of the high-desert community known as the pinyon-juniper woodland because of the pinyon pines and juniper shrubs that dominate the vegetation.

The trail ends at an overlook with breathtaking views into Deep Canyon, part of the Phillip L. Boyd Deep Canyon Desert Research Center. Scientists from across the nation come here to conduct field studies on desert plants and animals. From the Cahuilla Tewanet Overlook one can also view 5,141-foot Sheep Mountain to the east and Toro Peak, the highest point in the Santa Rosa Mountains (8,716 feet) to the south.

Persons wishing to spend a cool evening above the valley floor can do so at the Pinyon Flats Campground, the only public campground in either the Coachella Valley or the Monument. Located about two miles west of the Cahuilla Tewanet Overlook, the campground is situated amidst pinyon pines and junipers at 4,000 feet, slightly above the elevational limit of the Sonoran Desert.

Just three miles further up and west of the campground you'll be driving in the midst of the chaparral environment. This is a non-desert plant community dominated by tall, dense shrubs so closely packed that they form an impenetrable barrier to even the most persistent hiker. Many of the plants are adapted to recurring wildfires. They survive by resprouting

The Greater Roadrunner,
Geococcyx californianus, *can*   69
be seen crossing Highway 74.

after conflagrations or producing seeds that only germinate after intense heating. The dominant plant is ribbonwood (*Adenostoma sparsifolium*) but oaks (*Quercus cornelius-mulleri*), manzanita (*Arctostaphylos glauca*), and chamise (*Adenostoma fasciculatum*) are also common. As you drive through this area be on the lookout for mule deer (*Odocoileus hemionus*) crossing the highway.

Our palms to pines journey ends at the meeting place of highways 74 and 371, at an elevation of 4,919 feet, and the beginning of the yellow pine forest. Were you to continue west and north on Highway 74, in forty-five minutes you would arrive at the picturesque, mountain forest hamlet of Idyllwild.

To reach our final destination we have traveled a little more than twenty miles south and west of the Monument Visitor Center, and have risen almost a mile in elevation. Along the way we have driven through mountains, past canyons and palm oases, and seen four plant communities all in the space of one hour.

It sure beats a day-long drive through Kansas.

How to get there

Take Highway 111 to Palm Desert. Turn south (right if you are driving from Palm Springs; left if you're coming from Indio) onto Highway 74, the Palms To Pines Highway. Drive as far as your time and inclination take you but go at least 20 miles to the junction of highways 74 and 371.

Coachella Valley Preserve

What you will experience: *San Andreas Fault, native desert fan palms, desert springs, endangered desert pupfish*

By all rights the Coachella Valley Preserve shouldn't be. With its many desert springs, scenic vistas and miles and miles of flatlands, these thirty-thousand acres of desert landscape should have long ago been gobbled up by development, and converted to golf courses, housing tracts and shopping malls.

Instead, the preserve is much the way it has always been, with picturesque sand dunes, smoke tree-lined washes, flood-carved canyons and some of the largest desert fan palm oases in existence. With the exception of certain dune areas open only to naturalist-led tours, the entire Preserve can be visited on any day of the year.

How were these desert lands saved from development? The history of the Preserve is complex but its origin involved the first mayor of Palm Springs, a biologist and, of all things, a lizard.

The lizard was the Coachella Valley fringe-toed lizard (*Uma inornata*), today a threatened species, found only in areas of windblown sand in the Coachella Valley and nowhere else in the world. For decades the lizard received much notoriety from biologists because of the odd fringed scales on its toes and ability to escape from predators by dashing over dunes and disappearing headfirst into sand.

The biologist involved in the establishment of the Coachella Valley Preserve was Bill Mayhew, a professor of zoology at the University of California, Riverside. Mayhew had studied the lizard, was familiar with the region, and saw the lizard's habitat being rapidly lost to development.

The mayor in the emerging scenario was Phillip Boyd, a longtime resident of the Coachella Valley, the founder of the Living Desert and a benefactor to UC Riverside. Boyd, too, saw the continued influx of people into the Valley and wanted to preserve a part of the desert for posterity. Boyd and Mayhew were not directly responsible for the Coachella Valley Preserve, but did end up saving thousands of acres in what is now the

Dunes meet palms at
Thousand Palms Oasis

Boyd Deep Canyon Desert Research Center, south of Palm Desert. As the first individuals to bring to fruition a plan to preserve Coachella Valley landscapes, they planted a critically important seed for later land preservation.

Mayhew, however, wasn't satisfied with just Deep Canyon. In ever so subtle ways, he pushed the notion that all of the valley's habitat types should be preserved including the vanishing dunes in the valley center. When the U. S. Congress passed the Endangered Species Act in 1973 the stage was set. The lizard was listed as a threatened species in 1980 and the act required a habitat conservation plan be implemented to insure the lizard's home would be protected. With the task-oriented approach of the U.S. Fish & Wildlife Service, the timely fund-raising clout of the Nature Conservancy, the scrutiny of Mayhew, and the support of County Supervisor Corky Larson, the plan was enacted and in 1986 the Preserve was formed. Today it exists not just for the lizard, but for several rare, threatened or endangered species including a plant known as the Coachella Valley milk vetch, the flat-tailed horned lizard and an odd insect, the Coachella Valley giant sand treader cricket.

The focal point of the Preserve is Thousand Palms Oasis, a group of palm groves that, at my last count totaled several hundred, not one thousand, individual trees. Nonetheless, it certainly seems like there are a thousand of the stately trees. In several places the forty-foot palms are so dense that the sun is completely obscured. All of the palms belong to the species *Washingtonia filifera,* best known as the most massive palm species in North America. (These are the same kind of trees that line Palm Canyon Drive in downtown Palm Springs.) The skirt of dead fronds is the palm's most unique feature. It provides nesting and roosting sites for numerous flying animals including the rare and harmless southern yellow bat.

A recent land acquisition now connects the Preserve with Joshua Tree National Park. The connection allows plants and animals to move between the Preserve and the Park, facilitating genetic exchange. This will insure that plant and animal populations remain healthy and vigorous.

74

There is a visitor center located in the heart of the main grove of palms. It is a rustic house with walls made of all things, palm trunks. Volunteers staff the desk on most days and can help orient visitors. Bathrooms and tables are located in the oasis creating one of the most convenient and beautiful places to have a picnic anywhere in the Coachella Valley.

The valley's last sand dunes exist in the Preserve. They are closed to casual entry but visitors can sign up for guided nature walks that traverse the dunes. The walks are scheduled each spring. Contact the Preserve Manager at (760) 343-1234 for dates and times.

In scattered ponds around the oasis, and accessible by trails, one can see desert pupfish. Known affectionately to ichthyologists as *Cyprinodon macularius macularius*, these smaller-than-thumb-sized fish are restricted to a few isolated desert springs and waterholes scattered across the Sonoran Desert. They are so rare that the federal government has classified them as endangered and introduced them at several new sites in an attempt in insure their survival. The tiny ponds at Thousand Palms Oasis are among these refuges. The fish can be easily viewed in spring after they have awakened from winter hibernation and emerged from mud at the bottom of their pond.

A splendid, not-to-difficult hike starts from the oasis parking lot just off Thousand Palms Canyon Road. The trail heads mostly east towards the hills on the far side of a dry wash and more or less parallels the San Andreas Fault. Both trail and fault lead through a remarkable array of environments including an enormous wash dense with smoke trees and a creosote-dotted plateau. Hikers will see excellent views of Coachella Valley, and not one but five additional palm oases. More than any other single hike, I personally recommend this one for the best visual overview of the Coachella Valley. From the plateau one can see the Salton Sea to the southeast, the valley's and Preserve's last remaining dunes to the south, and the peaks and ridges of the new Santa Rosa and San Jacinto National Monument to the south and west. For more details on this hike grab a copy of Philip Ferranti's *120 Great Hikes in and near Palm Springs.*

Hairy Sand-verbena, Abronia villosa, *was once the Valley's most common wildflower*

The hike begins in Thousand Palms Wash, infamous among locals for the torrent it carried back in 1977. On the evening of September 10 of that year, a thunderstorm barreling northward from the Sea of Cortez slammed into the Little San Bernardino Mountains. As the moist air was drawn over the mountains a massive condensation of moisture occurred resulting in rainfall of near-biblical proportions. Some estimates placed the rainfall on the night of September 10 at four inches in a single hour. As the inevitable storm waters raced down the mountain canyons they deposited massive amounts of sand in the Colorado River Aqueduct, plugging it and forcing its flows down the wash channels. By morning, the results of what was to become Valley's most voluminous flash flood in recorded history were dramatically evident. Two-thirds of Thousand Palms Oasis was gouged out of the desert with more than a hundred palm trunks sent onto the valley floor, some more than six miles from where they grew. Millions of dollars in property damage was inflicted, most of it in the sparsely populated community of Thousand Palms. Some homes were lost, vehicles were buried in mud, and several roads were closed for weeks. Today, the only evidence of the flood is the smoke tree forest at the beginning of your hike. The trees are restricted in their distribution to places subject to flash floods.

Though the Coachella Valley Preserve was originally established to protect a species threatened with extinction, both visitors and residents are the real beneficiaries of its existence. As more and more land becomes occupied by homes and businesses, the Preserve insures the existence of a place where everyone can walk across the sand, hear only footsteps and experience the smells, sights and sounds of the desert.

How to get there

From the west, take I-10 to Ramon Rd. Turn left (east) on Ramon Road. Drive 4.5 miles and turn left (north) on Thousand Palms Canyon Road. Drive 1.5 miles and turn left into the visitor's parking area. From the east, take I-10 to Washington St. Turn right on Washington St. Drive 4.5 miles north and then west to Thousand Palms Canyon Road.

The endangered desert pupfish,
Cryprinodon macularis

The Salton Sea

What you will experience: An enormous desert lake, remarkable bird life including pelicans, snow geese, and herons, riparian habitats, shoreline tranquility

I was moving with the current and used my paddle only to push the canoe away from streamside vegetation. The cattails were so dense they formed a canopy over my head. It was quiet and I heard the muskrat before I saw it. Like a miniature beaver, it swam across the channel in front of the canoe. When it walked onto shore, I could see the long, hairless tail and the smaller size, immediately distinguishing it from a beaver. Further downstream an enormous grey bird leaped into the air from its hiding place and quickly flew out of sight. The white and black markings on its head indicated it was a great blue heron, the largest fish-eating bird in the region. Before reaching open water, the pointed nose of a soft-shell turtle emerged off the bow followed quickly by the eyes. When it saw me it dove for the bottom. After nearly two hours of moving through the narrow channel it was good to reach open water. Open water it was, for I had been carried onto California's largest lake.

What makes my canoeing experience odd is that I was smack in the middle of the Colorado Desert, a region second only to Death Valley as the hottest and driest place in California. Where I was canoeing, average annual precipitation barely reaches three inches and July maximum temperatures average a whopping 108 degrees Fahrenheit. With such meager precipitation and enormous evaporation rates, finding water here, much less canoeing, should have been as impossible as catching fish. The great blue heron, however, finds it easy to catch fish here. I was paddling in the Whitewater River Channel, a channel that normally carries runoff water coming from surrounding agricultural fields. Eventually it empties into the great body of water known as the Salton Sea.

At thirty-six miles long and fifteen miles wide, the Salton Sea has the largest surface area of any body of water in California. And there are more than ten different kinds of fish including corvina, the most popular catch among sports fishermen. Inevitably questions arise about the Sea. "How did such an enormous lake come to be?" "Why is it filled with so many fish?" "Is the Salton Sea a natural phenomenon?"

Cattails at Whitewater River/
Salton Sea junction

Today's Salton Sea, and the ecosystem that it supports, is the result of human meddling. Prior to 1905 it was a giant salt pan that filled with a few inches or feet of water several times each century. Heavy winter precipitation, together with runoff from the mountains surrounding the Imperial and Coachella valleys, caused these minor fillings. Typically, all the water evaporated away in a few days, weeks or months. Most of the time it was a dry, hard, alkaline-encrusted lakebed. All this changed in 1905. Decades before, government surveys noted that a great depression existed in the middle of the Colorado Desert. It was referred to as the Salton Sink and was situated below the level of the oceans. They also noted the Colorado River was above sea level and ran year-round unchecked into the Gulf of California. The river was less than one hundred miles away from the depression.

By the early 1900s entrepreneurs saw an opportunity to get rich by breaching the banks of the river and allowing the water to easily and cheaply flow downhill into the depression. Through a system of channels, the water would be used to irrigate crops that could grow year-round in the mild winter climate. It was projected to be a bonanza of land sales, water allotments, and agriculture. But in 1904 and 1905 snowfall was heavy in the Rocky Mountains and spring rains were warmer than usual resulting in rapid snow melt. The headwaters of the Colorado River filled with runoff and sent record-breaking river levels towards the Gulf of California. The canals and breaches in the riverbank, constructed to siphon off part of the flow, could not handle the high river levels. Within weeks the entire Colorado broke out of its normal course and took the easier route to the lower-lying depression. For nearly two years the river ran unchecked into the Salton Sink. Eventually the breach was closed by herculean efforts of Southern Pacific Railroad workers who dumped huge boulders into the breaches. The river was sent back into its proper channel but not before an enormous lake had formed, the Salton Sea. Today, the sea persists because of agricultural runoff water and still lies below the level of the oceans, about minus 228 feet. Maximum depth of the sea is approximately fifty feet.

As its name suggests, the Sea is salty, in fact about 30% saltier than the ocean. This comes as no surprise to geologists. The Sea lies in a great depression or, as it is technically known, a graben. There are more square miles of land below sea level here than anywhere else in the Western Hemisphere. Because of the low elevation, it is easier for water to flow into the sink than into the oceans, exactly what the Colorado River has done several times over a period of many thousands of years. The last

Salton Sea shoreline

natural inundation of the Salton Sink lasted for at least a century and created Ancient Lake Cahuilla, a body of water ten times larger than today's Salton Sea. The shoreline of this ancient lake can be seen as a light/dark interface at forty-two feet above sea level on the east faces of the Santa Rosa Mountains.

By the time the Spanish explorer Juan Bautista de Anza arrived in the region in 1774, the Colorado River once again had changed its course and returned its waters to the Gulf of California. The ancient lake had evaporated away and Anza saw only a dry, salt-encrusted playa.

With no outlet to the ocean, the Salton Sink accumulates not just water on occasion but salt as well. Sodium chloride and other types of salt are found in most rock formations to a greater or lesser degree. Rain erodes the rock, and any salt becomes suspended in the runoff. Eventually the runoff finds its way to the ocean or any depression along the way. In desert environments summer temperatures are hot, evaporation rates are high, and a lake in the depression is likely to disappear into thin air. The salt, however, does not evaporate and is left behind. Over eons of time salt accumulates, either in the remaining water or in the sediments on the dry lake bed. This is a natural recurring phenomenon that happens everywhere in dry climates and explains why the Great Salt Lake, the Red Sea and the Salton Sea are so salty. Any attempt at changing this natural process can be expected to be difficult, costly and require a never-ending commitment. Current investigations into preventing the Salton Sea from becoming even saltier must face this stark reality.

It should come as no surprise that creatures other than humans have discovered the Salton Sea. It is a body of water that supports lots of fish (most were introduced by the California Department of Fish & Game). The Salton Sea is also directly connected to the Colorado River via canals and lies not far from the Gulf of California. These latter two ecosystems lie along an important bird migration flyway. One would naturally expect that birds would eventually discover the sea.

The numbers and diversity of birds at the Salton Sea is quite extra-ordinary. To date, four hundred species representing well over a million individual birds can be expected in and around the Sea each year. This makes it one of the premier, if not the best, bird watching location in the entire country. Even a novice or unwilling spouse cannot help but be impressed with the Sea's bird life.

Below are my favorite three stops to make on a day-long trip to the Salton Sea.

The North Shore - *Starting in Indio head southeast on Highway 111 towards the town of Mecca. At Mecca, turn south onto Lincoln Street. This street dead ends at the Whitewater River Channel. Continue south up on the levee road (dirt). Drive slowly and stop before the road gets too sandy or muddy, otherwise you'll wish you had four-wheel-drive. Walk on the levee to the Sea from here. Great blue herons are always observable along the freshwater channel to your right. Also expect to see the beautiful white-feathered snowy egrets at the channel and the cute little ruddy ducks with their upturned tails out on the open sea.*

Salton Sea State Recreation Area – *Returning to Highway 111, turn right and continue heading southeast. A little more than ten miles from Mecca you will arrive at Parkside Drive. Turn right into the State Recreation Area. There are great views of the Sea at this location and a visitor's center that describes the history and ecology of the area. I usually see American white pelicans here as well as black-necked stilts and ring-billed gulls.*

Sonny Bono National Wildlife Refuge – *Birds abound in this area. In winter, expect hoards of Canada and snow geese as well as American avocets. Of special interest are the bubbling mud pots that you can approach on foot. To reach the refuge area and mud pots continue on Highway 111 and pass through the town of Niland. Turn left (west) on Sinclair Road. Continue 1.5 miles to English Road and turn right. Continue another 1.5 miles to Schrimpf Road and turn left (west). Drive two miles to Davis Road and park you car at the roadside. The mud pots are to your right, not more than twenty-five yards northeast of the intersection of Davis and Schrimpf roads.*

On the way home, I recommend stopping at **Oasis Date Gardens** *for a yummy date shake. It's on the west side (left) of Highway 111 just south of Avenue 58. I have one every time I go to the Salton Sea.*

Canada Goose,
Branta canadensis

Painted Canyon

What you will experience: *Colorful fault-contorted rocks, desert trees, vertical canyon walls, breath-taking vistas*

If you want to touch the San Andreas Fault, view folded rock, walk in shoulder-width canyons and discover wood that sinks, you must drive to Painted Canyon. This is one of the most unique destinations in the Coachella Valley and perfect for a one-day, family outing.

The adventure begins when visitors turn northwest off State Highway 195 near Mecca and onto Painted Canyon Road. Along the way are several interesting Sonoran Desert plants, most notably the endemic ironwood tree and the strange-looking ocotillo. The ironwood tree is known among desert aficionados as the plant whose wood is so dense that hacking at the trunk quickly dulls the sharpest ax. The wood is also heavy and sinks when placed in water. Look for the ironwood's pink blossoms in May. The ocotillo is the Colorado Desert's strangest-looking plant. Dozens of unbranched stems form an inverted cone that resemble the frame of a teepee turned upside down. Providing winter rains have been ample, expect the stems to be covered with green leaves in March and tipped with flaming-red flowers in April. Both of these plants thrive in California's low, hot Colorado Desert where their sensitive tissues rarely need cope with subfreezing temperatures.

As the unpaved road veers to the north, the uniqueness of Painted Canyon becomes obvious. All around are sedimentary rocks, identified by the conspicuous layering of sediments that eons ago were deposited by water. Normally, such layers would be aligned horizontally, one layer upon the other—the kind of stratification that characterizes the Grand Canyon. But recent geological activity along the San Andreas Fault has radically altered the stratification here. Near the canyon mouth, lines of ancient sediments have been bent upward, downward, and even folded on top of each other. The contorted layers are solid rock, but from their appearance it would seem that they are as malleable as taffy in a candy-makers kettle.

The fault itself lies along the base of the Mecca Hills and crosses the entrance to Painted Canyon. Visitors travel over the fault at a right angle as they drive into the canyon. Contrary to popular myth, the fault is not a

Fault-tilted rock layers and smoke trees in Painted Canyon

line in the sand but a fracture zone varying in width from several dozen yards to more than a mile. In the canyon, the first evidence of the fault is horizontal rock layers that have been turned vertically. They are unmistakable on your left after entering the canyon. More evidence is on the right where two different rock types have been shoved up against one another along an enormous crack. A final piece of fault evidence occurs just outside Painted Canyon. Rising groundwater has seeped up through the fault and provided water for two small palm oases, Sheep Hole Palms and Hidden Palms. They can be located on topographic maps.

Painted Canyon and the Mecca Hills (of which the canyon is part) both straddle and ultimately are the result of movements along the San Andreas Fault. The movement occurs when pressure from the earth's molten interior jostles pieces of the earth's crust. These pieces, known as plates, more or less rest on the molten material. The San Andreas Fault is the meeting place of two of the plates: the North American Plate on the northeast and the Pacific Plate on the southwest. Though most movement has been horizontal, when these enormous pieces of the earth's crust move past each other some bending and compression occurs. It is this compression that has bent and folded the earth resulting in the hills, vertical walls, and mountains that we call the Mecca Hills. The hills themselves were lifted above the surrounding terrain beginning about one million years ago. Runoff from the Little San Bernardino Mountains to the north began carving Painted Canyon at the same time but particularly during the ice age that did not end until ten thousand years ago. (Though this may sound like a long time ago, to a geologist contemplating a 4.5 billion-year-old earth, anything that happens in the last few million years is considered a recent event.)

The colors of Painted Canyon are nearly as impressive as the contorted rock walls. Cliff faces can be muted beiges, tans, and browns, quiet yellows, and in places even pale green. Oranges and reds are also present and some rocks are nearly black. Again, the variety is attributable to fault movements that have brought together a collage of sediments including marine, river and desert wash materials and then subjected them to pressure, heat, steam, and percolation. According to valley geologist Harry Quinn, most of the colors reflect iron minerals that have combined with oxygen and formed various shades of yellow, orange, red and brown.

In some places iron compounds have lost oxygen (known as reduction) resulting in rocks with a greenish cast. Some of the darkest colors often

Blue Palo Verde,
Cercidium floridum,
in bloom

indicate the presence of manganese. In all of the Coachella Valley, there is no other site that yields such a variety of rock color.

Driving north further into the canyon, visitors will notice that both sides of the road are dotted with small, blue-gray trees. These are smoke trees, a name derived from their resemblance to a puff of smoke, at least from a distance. More shrub than tree, this member of the Pea Family is the great soothsayer of flash floods. Smoke tree seeds germinate after tumbling about in floodwaters. As the floodwaters recede, the seeds absorb moisture, germinate, and quickly send a taproot several feet into the sand. (The above-ground portion of the seedling is likely to be less than five inches in height!) Smoke trees cannot survive on local rainfall alone and are thus born in, and restricted to, dry washes where floodwaters can be expected at least a few times each decade. Thus, if one contemplates purchasing land for a desert home, beware of even one smoke tree on the property—it is a good predictor of flooding.

As you continue driving up the canyon the oxidized canyon walls become closer, taller, darker and older. About four miles on Painted Canyon Road, the sand gets softer and a sign warns that the road is not maintained ahead. Do not proceed further even if you have a vehicle equipped with four-wheel-drive. The canyon forks here, but the best walks and picture-taking are to the right. Most persons will be content to walk up the main canyon to a nearly impassable dry waterfall about a mile from your parked car. Along the way the canyon continues to narrow, and the pink vertical walls eventually reach more than two hundred feet in height.

The more adventuresome may wish to hike through Ladder Canyon, a slot canyon that requires one to turn sideways to get through several narrow places. The entrance to Ladder Canyon is also via Painted Canyon, about one-quarter-mile from where cars are parked at the canyon divide. Look for a sign that directs hikers to the left, up the right side of a rock slide. The entrance is hidden by the slide. Once in this side tributary you'll discover why they call it Ladder Canyon. There are several ladders, usually well maintained, that assist hikers in reaching the upper levels of the canyon. Test them carefully before you proceed.

Painted Canyon is managed by the Bureau of Land Management (BLM) and, so far, it continues to be open to the public without restriction. With its fascinating geology, scenic splendor, and accessibility, it's comforting to know that the canyon is forever on public land.

How to get there

Painted Canyon is situated at the southeastern end of the Coachella Valley, near the town of Mecca. To get there from Palm Springs take Interstate 10 to Highway 111 and continue driving southeast until you reach State Highway 195 (66th Avenue or Box Canyon Road). Turn left (east) and drive about five miles through the town of Mecca, past numerous vineyards and citrus orchards, and over the All American Canal. Once past the canal look for a road sign that will direct you onto a four-mile, unpaved road that heads northwest then turns north into Painted Canyon. (The road sign is not always present so remember that Painted Canyon Road is the first graded dirt road to the left after the canal.) Although a four-wheel-drive vehicle is recommended on this road a standard passenger car can usually make it. Do not stray from the road as the shoulders can be sandy and soft and it is easy to get stuck. Park where you find the "Road Not Maintained" signs.

The region is riddled with
narrow side canyons

Joshua Tree National Park

What you will experience: *Forests of Joshua trees, dense stands of cactus, palm oases, picturesque boulder fields, the best view of the Coachella Valley*

Let's say your distant cousin from England was coming to America, to the Coachella Valley, for the first time. She expressed an interest in seeing many attractions but had reserved one full day to experience the desert. She wanted to see the clear blue skies, the unending vistas, and the muted brown and purple mountains she had read about. A few strange-looking plants like cacti would be nice and maybe even a rattlesnake from a safe distance. Your cousin wanted to experience all that was different, everything she would never find in England. Where would you take her if there was only one day to experience the California deserts?

Ask me this question and I'll suggest a drive to Joshua Tree National Park. In one day you will experience not one, but two of California's desert regions—the Sonoran and Mojave. You will also see enough cacti, yuccas, lavender-colored mountains and bright blue sky to satiate the most demanding traveler.

Joshua Tree National Park holds all these wonders and much, much more in its 800,000 acres of desert landscapes. (The Park is nearly as large as Rhode Island!) To see a good part of it in a day I recommend departing no later than 8:00 a.m. Otherwise, you'll be racing by some of the choicest spots. Take Interstate 10 east to the big highway sign that appropriately says "Joshua Tree National Park." The turnoff is about a 45-minute drive from Palm Springs.

If you're departing anytime from mid-February to mid-May, the action starts as soon as you turn off the freeway and head north up the grade. You're driving on a Sonoran Desert bajada formed by sheet flooding over many millions of years. The bajada slopes to the south, gets more solar radiation than other areas and is consequently one of the first places to warm up in late winter. In years of average or above-average winter precipitation this area is an outstanding place to find wildflowers. Here one can see fields of desert gold poppies, two kinds of lupine (the lavender Arizona lupine and the blue Coulter's lupine), desert dandelion and

The Joshua tree is the nation's largest yucca.

The Cholla Garden in
Joshua Tree National Park

several dozen additional species. My personal favorite is a red-flowered shrub known as the chuparosa, the Spanish word for hummingbird. The chuparosa is becoming increasingly scarce in the Coachella Valley as a result of the development of land at hillside bases, the chuparosa's favorite haunt. Fortunately, the species is protected forever in the Park.

Still heading north, the next stop is Cottonwood Spring. To get to the spring you'll need to turn right at the sign and make a quick stop at the Visitor Center. Pay the entrance fee here, buy your wildflower guides (Jon Stewart's *Colorado Desert Wildflowers* and *Mojave Desert Wildflowers* are best) and take advantage of the last flush toilets you'll see the rest of the day. A short (less than one mile) drive takes you right to Cottonwood Spring.

Water in the desert is always a surprise. By all rights there shouldn't be any since dryness is what defines a desert. That is why Cottonwood Springs is so out of context. First there are the cottonwood trees that only grow in perpetually moist soil. At Cottonwood Spring these share space with moisture-loving desert fan palms. Together the trees form a miniature rain forest complete with canopy, shaded grottos, and colorful singing birds. Only the monkeys are missing in this decidedly undesertlike place.

Visitors can spend a half-day here watching and listening for hooded orioles (*Icterus cucullatus*) and other oasis birds. But Pinto Basin, the strange ocotillo, and the teddy-bear cholla are not far away.

The road leaving the Cottonwood Spring Visitor Center briefly heads up and north before plummeting down into a great basin. This is Pinto Basin, one of the most isolated places in the Park and in all of the California deserts south of Death Valley. At night the only light one sees comes from the stars save for an occasional wayward motorist. I have always likened this region to the Coachella Valley when the Cahuilla Indians had it all to themselves.

It is in the Pinto Basin where one finds an area rich in ocotillos. This is the northwestern limit of the species' range as all other wild ocotillos are found to the south or east. Besides the strange upside-down look of all ocotillos, there is something else odd about the ones in Pinto Basin. They are big, really big. I confess that I have never actually taken a tape measure and determined the exact height and circumference of some of

these monster ocotillos, but you'll certainly never find any this size in Arizona's Saguaro National Park or Organ Pipe Cactus National Monument. Botanists would expect to encounter small, struggling specimens at the limit of a species range, not huge, stellar individuals. As yet, there is no explanation for this mystery.

Just east of the ocotillo patch, and along the road on which you have been traveling, there is an even more formidable plant. It grows in such numbers and under such crowded conditions that it is reminiscent of a vast infantry regiment. The plant is the jumping cholla, *Opuntia bigelovii*. As a member of Cactus Family, no one should be surprised that the jumping cholla has spines. What is surprising is the number and density of them. This has to be the spiniest of all cacti. If you took a soda straw you could not poke the green skin of the cactus itself. In fact you can hardly see it.

The spines protect the cactus from being eaten by hungry herbivores and shield it from intense solar radiation. They also serve another function. The spines assist the jumping cholla in traveling around, or being dispersed as biologists would say. The spines of the jumping cholla are not only very sharp, they are covered with backward curved flanges that flare out like the barbs of a harpoon after penetration. They are extremely difficult to remove once they have entered the flesh of a jackrabbit, coyote or human. Since the spines also stay attached to a segment of the cholla, that means a piece of the cactus is going for a ride of a few feet or a mile before it drops or is pulled off. It is this lemon-sized segment, not the cholla's seeds, that are responsible for the plants reproduction. Once it falls to the ground the segment produces a root, affixes itself to the soil and grows a new jumping cholla.

Leaving the cholla parking area our journey continues upward to the very highest reaches of the Park. Heading north and west we soon reach the four-thousand-foot level and the beginnings of the Park's namesake, the Joshua tree, *Yucca brevifolia*. Though not the largest species of yucca on our planet, it is certainly the largest in America. At least a few individuals have been found that reached more than fifty-feet in height with trunks more than four feet in diameter. The Joshua tree is the only tree that survives on the open desert flatlands, or put another way, the only desert tree that is not associated with arroyos. It also marks our departure from the "low" Sonoran Desert and our entry into the "high" or Mojave Desert.

In this western area of the Park, Joshua trees are not the only objects that noticeably protrude above the landscape. There are an equal number of Mojave yuccas (*Yucca schidigera*) and a greater number of giant, isolated, granite outcrops. These huge boulder areas have descriptive names such as Jumbo Rocks, Cap Rock and Wonderland of Rocks. The enormous stone massifs are erosional remnants, harder rocks that withstood the effects of temperature extremes, wind, and water erosion much better than the softer material eroded away around them. What we see today is an assortment of giant boulders, some alone, some piled one upon another with sizes varying from a few feet to over a hundred feet in height. Their rounded edges, alcoves and miniature caves beg to be explored and climbed upon by visitors of all ages. Together with the Joshua trees, the boulder fields in Hidden and Lost Horse valleys are the best locations in the Park for photography.

We continue west and then south on the paved roadway where our journey ends at Keys View, a five-thousand-foot-plus scenic overlook with the very best view of the Coachella Valley. Temperatures are always cool here, even in July. Alongside scattered junipers and pinyon pines one can observe the Salton Sea, California's largest lake; the Salton Trough, the greatest area of below sea level terrain in the Western Hemisphere; and the northeast face of the San Jacinto Mountains, the steepest escarpment anywhere in North America.

As you drive home through the towns of Joshua Tree and Yucca Valley, it is easy to be overwhelmed by the diversity of landscapes one encounters in a single day. Just remind yourself that this is California, a land that is continually reconfigured by forces deep within the earth. Wonders can be found everywhere.

How to get there

From Indio, take I-10 22 miles east to the Joshua Tree National Park turnoff. Turn left (north). Stay on this main Park roadway for 37 miles then turn left at first paved intersection. Continue for 10.4 miles then turn left again and drive 5.5 miles to Keys View. I recommend leaving the Park through the town of Joshua Tree. Pay entrance fees and obtain a Park map at the Cottonwood Visitor Center, encountered shortly after entering the Park.

Ocotillo,
Fouqueria splendens,
in Pinto Basin

Berdoo Canyon

What you will experience: *High and low deserts in an hour, remote back country, billion-year-old rocks, solitude, spring wildflower diversity*

Joshua Tree National Park is the Coachella Valley's nearest neighbor. That's why I was surprised when I was told by a long-time desert resident that "You can't get there from here."

It was many years ago and, as a newcomer, I was anxious to discover the new surroundings. The acquaintance was describing how, in order to visit Joshua Tree National Park, one must leave the Coachella Valley and enter through the high desert town of Joshua Tree. Alternatively, one could drive east out of the valley on Interstate 10, and enter at Cottonwood Springs. In short, you couldn't get to the Park directly from the Valley.

It wasn't until a year later that I discovered my informant was incorrect. You can get there from here. Hikers and drivers may enter Joshua Tree National Park without ever leaving the Coachella Valley.

A bit of geography is necessary at this point. No doubt surprising to most Coachella Valley residents is the fact that two-thirds of the mountains one sees when looking north are part of the Park—from the peaks and ridges right down to the valley floor. At the mouth of every canyon leading up into the Park (and into the Little San Bernardino Mountains), is a small, official-looking sign with the words "Park Boundary." These words clearly delineate who is in charge of the land. The confusion surrounding the Park's proximity lies in the fact that those four- by ten-inch boundary signs are just about the only direct evidence that the Park and the valley are joined at the hip.

Unlike the south side of the Coachella Valley where nearly every canyon is either closed to hikers or has restricted access, the north side is open to all. Not one canyon is closed or has restricted access. In short, you can walk directly into the Park from the Coachella Valley, at any time and in many places.

A four-wheel-drive vehicle negotiates Berdoo Canyon

You may also drive into the Park from the Valley. This is not one of the loopy routes where one spends as much time driving away from the Park as towards it. This is a direct route leading from the low Colorado Desert of the Coachella Valley to the high Mojave Desert of Joshua National Park. After driving just two miles you're in.

There is a catch as to what valley residents should already be aware of. The catch is that the road is unpaved, unkept and, at times, nearly impassable. It is so bad that the Southern California Automobile Club has removed the road from its latest Riverside County map. Traveling up it without a high-clearance, four-wheel-drive vehicle is like hiking barefoot—sooner or later not having the right equipment is going to get you in trouble. With such a vehicle, however, one can experience the best, legally used four-wheel-drive road in the entire region. (At this point I probably should mention that I have driven this route next to a hundred times and have always made it through without vehicle damage of any kind.)

The route begins off Dillon Road in the southeastern end of the Coachella Valley and north of the city of Indio. For those planning the trip, the turn off Dillon Road and onto the road into the Park is eight and one-half miles east of Thousand Palms Canyon Road and nearly six and one-half miles west of Interstate 10. Years ago there was an official sign marking the turn but it was eventually obliterated by target shooters.

Turning northwards, drivers will be pleased to discover that the first mile or so of roadway is paved. The road runs on top of an ancient pile of sand, gravel and rock carved out of the canyon by flash floodwaters over the past twenty million years. This miles-wide pile of erosional material is called an alluvial fan owing to its fan-like shape. Alluvial fans occur wherever canyons open onto valley floors. It is a natural process of erosion best observed in deserts where vegetation doesn't hide the earth. On the way up the fan you'll also see very recently deposited debris in the form of rubbish. These trash piles are obviously the result of unthinking valley residents. (It should be noted that the illegal dumping problem was seriously exacerbated when Riverside County Supervisors voted to restrict the public's access to legal dump sites and charge a fee for each use.)

The pavement ends as the road dives to the canyon bottom. This is Berdoo Canyon, and the road stays tight on the canyon floor for six miles while slowly rising 3,500 feet in elevation. At 4,200 feet the canyon opens into Pleasant Valley in Joshua Tree National Park. No one knows precisely how the canyon got its name. My best guess is that it was named by one

Desert Chicory,
Rafinesquia neomexicana,
in Berdoo Canyon

of the miners that worked the canyon walls for minerals in the early 1900s. Berdoo seems to be a nickname for the Little San Bernardino Mountains.

Take time to enjoy the journey through Berdoo Canyon. Along the way you will see the region's most ancient surfaces. Some of the metamorphic rocks exposed in the canyon walls have been around for 1.2 billion years, long before giant mammoths walked the valley floor, before dinosaurs lived on Earth, and even before the first backboned creatures crawled onto land.

Berdoo Canyon cuts through the Little San Bernardino Mountains, the latter so named because they are not even half as high as the (big) San Bernardino Mountains to the west. Like their taller brothers, they were first born with the formation of the San Andreas Fault about twenty million years ago. Though the overwhelming majority of movement along the fault has been horizontal, some has been vertical. All the mountains surrounding the Coachella Valley are the result of this vertical movement. Berdoo Canyon has been forming ever since the creation of the fault and uplift of the mountains, carved with decreasing intensity by storm runoff and floodwaters.

Observant drivers may catch a glimpse of bighorn sheep. These magnificent, horned ungulates regularly traverse the region's canyons and mountainsides in search of food. If seen in the fall, rams may be searching for females. Unlike bighorn living in the mountains on the south side of the Coachella Valley, bighorn in Joshua Tree National Park have maintained stable populations for decades and may be the healthiest populations in the state.

It was in Berdoo Canyon that I once discovered a most unusual reptile. It was a snake, a relative of the boa constrictor known as the rosy boa (*Lichanura trivirgata*). This three-foot-long serpent has achieved near-celebrity status among the snake-fancying public because of its giant relative (the boa constrictor), a sloth-like rate of movement, and absolute unwillingness to bite when handled. (This is usually the species of snake used in classrooms to demonstrate how the most serpents are quite harmless.)

In the Coachella Valley, rosy boas are gray with drab reddish-brown stripes that are hardly discernible from the background. In short they're rather plain in appearance. The one I encountered in Berdoo Canyon, however, was truly beautiful. The stripes were distinctly orange and the ground color was a light tan. The hues were more reminiscent of a rainbow than a reptile. That was the only rosy I have ever seen in the canyon and I can't help but wonder if all of the resident boas are equally as beautiful.

Berdoo Canyon is a great place to botanize each spring, even if winter rains have not been bountiful on the valley floor. Mountains and their canyons usually receive more precipitation than surrounding lowlands and Berdoo Canyon and the Little San Bernardinos are no exception. What's more, runoff from storms brings even more water to the canyon bottom. Expect to see multitudes of yellow-flowered desert dandelions, white desert chicory and purple chia every spring. At the top of the road, in Joshua Tree National Park, grow hoards of Mojave yuccas, some of the densest concentrations I have ever seen. Look for enormous numbers of their erect, cream-colored flower clusters.

There is much to see during the drive up Berdoo Canyon but it is what one does not see that makes the trip most alluring. Even on a weekend day a driver or hiker may not encounter another person. In the canyon one is alone, cut off from the outside world without the use of a cellular phone. As the Park boundary is crossed, be prepared for a true wilderness experience, one that is unrivaled in the region.

How to get there

From Indio at Interstate 10, take Dillon Road 5.7 miles north and then west to Berdoo Canyon Road (not marked). Turn right (north) and drive up the alluvial fan and then into the canyon. A slow, hour drive up the canyon leads to Pleasant Valley and Joshua Tree National Park. The road is not maintained and may occasionally be blocked due to flash floods. Four-wheel-drive vehicles are recommended and high-clearance vehicles are required.

Big Morongo Canyon Preserve

What you will experience: Remarkable diversity of bird life, dense groves of giant cottonwood trees, accessible boardwalk through oasis heart, abundant shade and relatively cool temperatures in summer

It was Friday and I told my children that the next day we were going to see Big Morongo.

"Dad, what is Big Morongo?" my oldest daughter asked.

Fearful that they would anticipate a day of boredom and not want to go if I told them the truth I said "Big Morongo is a giant frog, the largest frog in the world."

Immediately my oldest daughter became suspicious. "How big is it?" she asked.

"No one has actually seen it, but many people have found its tracks and heard its roar at night. We're going to try and be the first people to actually see it."

"Is it scary?" my four-year-old asked.

"I don't think so. Probably not. Besides you're with me and I will protect you," I said while standing taller than my frame allowed.

I don't think my oldest daughter believed me. She was, however, still at an age when electing not to know was acceptable and it certainly made life more exciting.

The next morning my children were waiting to embark upon the adventure along with six other boys and girls on our block, each of whom, unbeknownst to me, were invited to search for Big Morongo.

That was many years ago and to this day no one has ever found the giant frog. No one ever will, of course, because Big Morongo is an ecological preserve, not an enormous amphibian. The concocted story was designed

The shaded, boardwalk trail that meanders through the Preserve

to entice my children, and inadvertently the neighbors', into visiting Big Morongo Canyon Preserve. We didn't see a huge frog but we saw many small ones, as well as a hungry looking coyote, a long-eared owl, and a giant tarantula (a relative of true spiders). Our adventure was the talk of the neighborhood for at least a week. The next day the children wanted to know if we could go look for Big Morongo again.

Big Morongo Canyon Preserve is found adjacent the small community of Morongo Valley off Highway 62, also known as the Twentynine Palms Highway. The area is high desert, the beginning of the Mojave, and usually ten degrees cooler than the Coachella Valley. It is one-half hour north of Palm Springs and an hour northwest of La Quinta; an easy drive from everywhere in the Coachella Valley.

But why drive to Big Morongo Canyon Preserve if there are no giant frogs? The reasons are many and include a lush desert oasis, the area's largest and most magnificent cottonwood trees and the best place in the region to see the beautiful vermilion flycatcher. I guarantee raptor sightings as this is one of the most productive bird-watching spots in the West. Early morning visitors have a chance of glimpsing a bighorn or bobcat. You may add to all this a gift box of amenities—shaded parking (beneath cottonwood trees), superb hiking trails, picnic tables, benches, real bathrooms, and a boardwalk (wheelchair friendly) that meanders through the heart of the preserve oasis. All these things make for a complete and comfortable nature experience for anyone of any ability.

I must confess that it is the huge cottonwoods trees that bring me back to the Preserve again and again. There are no larger cottonwood trees in the entire region and some reach more than eighty feet in height. They are Fremont cottonwoods, a member of the Willow Family and technically known as *Populus fremontii*. (The species is named after John C. Fremont, an explorer and one of California's first United States Senators.) It is usually this species that grows at remote desert springs, such as the one at Big Morongo. Cottonwood seeds are small and have numerous long, silky hairs that capture the wind. They can easily be carried aloft and transported vast distances before falling back to earth. Over time, at least one Fremont cottonwood seed has reached nearly every desert spring and waterhole in the desert regions of California.

As residents of temperate regions might expect, cottonwoods are trees that change with the seasons. At Morongo the trees are leafless in winter, luscious green in summer with leaves turning yellow to pale orange in

*Cottonwood and willow forest
in Big Morongo Canyon*

fall. Winter may not be the best time to admire cottonwoods but it is easy to see the hawk nests built the previous spring. One can also see dead branches and broken trunks, critically important to birds that build their nests in hollow tree cavities. Only in dead trunks and branches can woodpeckers excavate nests. Later, abandoned woodpecker nests are used by other, cavity nesting birds such as bluebirds, screech owls and titmice. These and many other cavity nesting bird species keep a lid on the number of insects and rodents in the environment. Without the dead branches and standing trunks there are no nest sites, fewer birds, and no checks on insect and rodent populations.

With its flowing stream and year-round supply of water it should come as no surprise that Big Morongo Canyon Preserve has an array of animal species that naturalists don't expect to find in a desert. Raccoons roam throughout the Preserve at night and are occasionally seen during daylight hours. Treefrogs are also active at night but are so amorous during early spring that they cannot contain themselves and are heard croaking their love song in broad daylight along the boardwalk. A strange reptile, known as the legless lizard (*Anniella pulchra*), is a denizen of cooler, moister coastal California yet here it is at Morongo. It looks like a small snake because it has no limbs. The presence of eyelids, however, classifies it as a lizard. Don't expect to see the legless lizard. It is fossorial, spending its entire lifetime burrowing through the soil.

As interesting as it may be to see fleeting mammals and secretive reptiles, most nature enthusiasts come to the Preserve to watch birds—and here at Morongo they are abundant, colorful and diverse. With its reliable water and multitudes of trees, the Preserve is the perfect desert outpost and an important stopover for migrants. I once drove up to Morongo when the Preserve Host called because he thought I might like to see three hundred migrating turkey vultures (*Cathartes aura*) roost in the cottonwoods for the evening. Most bird watchers would rather see the brilliant vermilion flycatcher, a species whose red plumage (male only) impresses even an uncooperative spouse. Dee and Betty Zeller, the Preserve naturalists, told me that sometimes these flycatchers arrive as early as March. A large, female great horned owl also catches the attention of casual visitors as does a red-tailed or Cooper's hawk. In all, 246 bird species have been recorded at the Preserve, more species than have been recorded in all of the Coachella Valley excluding the Salton Sea.

Recently, a new plant species has, on its own, entered the oasis environment at the Preserve. In the fall of 2005, I noted the appearance of several young desert fan palms, the kind found in the Indian Canyons and along the San Andreas Fault. I don't recall seeing them before. In the past it was thought to be too cold in Morongo Valley for fan palms. Apparently, things are warming up here. Given time, and an absence of freezing winter nights, the warm-weather palms may transform the cottonwood forests into palm groves.

How to get there

From Palm Springs take Indian Canyon Drive north to Interstate 10. Turn left (west) onto I-10 and drive 3.7 miles to Highway 62. Veer right and head north on Highway 62. Continue on Highway 62 for 9 miles and into the town of Morongo Valley. Turn right (east) on East Drive. Drive one block to entrance sign and turn left into the Preserve. Parking is on the left, beneath the cottonwood trees.

For more information about Big Morongo Canyon Preserve, including facilities and directions, go to their website at www. bigmorongo.org

Barn Owl,
Tyto alba

Suggested Reading

Much of the information found in this guide was taken from the publications listed below. Each is an excellent source for information on the natural history of the Coachella Valley region. Although some are available for sale in local bookstores, visitor centers, and interpretive institutions, all are available at one or more public libraries. The libraries at College of The Desert in Palm Desert, the Rancho Mirage Public Library and the Palm Springs Library Center have particularly good collections of local history and natural history books.

Bailey, V. J. 2003. *California Desert Resort Cities.* Desert Springs Publishing, La Quinta, California.

Baldwin, B. G. and others. 2002. *The Jepson Desert Manual.* University of California Press, Berkeley, California.

Bean, L. J. 1972. *Mukat's People.* University of California Press, Berkeley, California.

Bean, L. J. and K. S. Saubel. 1972. *Temalpakh: Cahuilla Indian Knowledge and Usage of Plants.* Malki Museum Press, Banning, California.

Chase, J. S. 1919. *California Desert Trails.* Houghton Mifflion Company, Boston, Massachusetts.

Cornett, J.W. 1989. *Desert Palm Oasis.* Nature Trails Press, Palm Springs, California.

Cornett, J. W. 2004. *Venomous Animals of the California Deserts.* Nature Trails Press, Palm Springs, California.

Ferranti, P. 2002. *120 Great Hikes In and Near Palm Springs.* Westcliffe Publishers, Englewood, Colorado.

Fleming, J. A. 1994. *Coachella Valley Hiking Guide.* Nature Trails Press, Palm Springs, California.

Glenn, T. 1992. *Birds of The Coachella Valley: A Checklist.* Nature Trails Press, Palm Springs, California.

Henderson, R. 1968. *Sun, Sand and Solitude.* Westernlore Press, Los Angeles, California.

Jaeger, E. C. 1957. *The North American Deserts.* Stanford University Press, Stanford, California.

Desert Agave,
Agave deserti,

Jaeger, E. C. 1965. *The California Deserts.* Stanford University Press, Stanford, California.

Jaeger, E. C. 1969. *Desert Wild Flowers.* Stanford University Press, Stanford, California.

James. G. W. 1906. *The Wonders of the Colorado Desert.* Little, Brown, and Company, Boston, Massachusetts. (two volumes)

James, H. C. 1960. *The Cahuilla Indians.* Malki Museum Press, Banning, California.

Massey, B. W. and R. Zembal. 2002. *Guide to Birds of the Salton Sea.* Arizona-Sonora Desert Museum Press, Tucson, Arizona.

Miller, A. H. and R. C. Stebbins. 1964. *The Lives of Desert Animals in Joshuua Tree National Monument.* University of California Press, Berkeley, California.

Norris, R. M. and R. W. Webb. 1990. *Geology of California.* John Wiley & Sons, New York, New York.

Ryan, R. M. 1968. *Mammals of Deep Canyon.* The Desert Museum, Palm Springs, California.

Sharp, R. P. 1994. *A Field Guide to Southern California.* Kendall/Hunt Publishing Company, Dubuque, Iowa.

Sharp, R. P. and A. F. Glazner. 1993. *Geology Underfoot in Southern California.* Mountain Press Publishing Company, Missoula, Montana.

Stewart, J. M. 1993. *Colorado Desert Wildflowers.* Jon Stewart Photography, available from Nature Trails Press, Palm Springs, California.

Ting, I. P. and B. Jennings. 1976. *Deep Canyon, a Desert Wilderness for Science.* Philip L. Boyd Deep Canyon Desert Research Center, Palm Desert, California.

Trent, D. D. and R. W. Hazlett. 2002. *Joshua Tree National Park Geology.* Joshua Tree National Park Association, Twentynine Palms, California.

Weathers, W. W. 1983. *Birds of Southern California's Deep Canyon.* University of California Press, Berkeley, California.

Zabriskie, J. G. 1979. *Plants of Deep Canyon.* University of California, Riverside, California.

Field of Arizona lupine,
Lupinus arizonicus

Where To Learn More

The Coachella Valley has many visitor centers and institutions where you can learn more about the geology, plants, animals as well as the first people to occupy the region.

Living Desert

Located in Palm Desert, take Highway 111 to Portola Avenue and turn south. Drive 1.5 miles to the entrance. View hundreds of living plants and animals native to the Coachella Valley region. Dozens of interpretive kiosks explain the natural history of the area. Progams, gift shop and cafe complete your experience. Admission fee discounts for children and seniors. Go to www.livingdesert.org for more information.

Agua Caliente Cultural Museum

Located in downtown Palm Springs, this museum is dedicated to the Agua Calient Band of Cahuilla Indians, the first human occupants of the Coachella Valley. Changing exhibitions, lectures and classes tell the story of these first people, where they have been and their accomplishments today. Go to www.accmusuem.org for more information.

National Monument Visitor Center

Interpretive exhibits describe what you can see and do while visiting Santa Rosa & San Jacinto Mountains National Monument. The Bobcat Bookstore provides books, maps and trail guides. A botanical walk around the grounds introduces visitors to native plants. Located in Palm Desert on Highway 74 at the mountain. Go to www.monumentaltreasure.com/points for more information.

Joshua Tree National Park Visitor Centers

Visitor centers are located at each of the entrances to the Park: Yucca Valley, Joshua Tree, Twentynine Palms and Cottonwood Spring near the south entrance. Exhibits at each center depict different parts of the Park's story. Rangers offer lectures and guided tours during the season. Go to www.nps.gov/jotr or www.joshuatree.org for more information.

Salton Sea State Recreation Area Visitor Center

Informative exhibitions describe the history and natural history of the Salton Sea, California's largest lake. The Visitor Center is located 30 miles south of Indio on Highway 111, near the community of North Shore. Go to www.saltonsea.statepark.org for more information.

Palm Springs Visitors Information and Reservation Center

Located at 2901 North Palm Canyon Drive in Palm Springs, visitors will find information on attractions, accommodations, and restaurants in the Palm Springs area. Guide books and informational literature are available and personnel are available to answer questions and make hotel reservations. Call 760-778-8418 for more information.

Palm Desert Visitor Center

Located at 72-567 Highway 111 in Palm Desert, this new facility offers information on attractions and places to stay while in the Palm Desert area. Books and maps are on sale and personnel are always available to answer questions. Nearby is the **Palm Desert Historical Society Museum** at 72-861 El Paseo. Go to www.palm-desert.org/visiting_visitorcenter.asp for more information.

Oasis Date Gardens

Dates are an important commercial crop in the region and the Oasis Date Gardens are the best place to buy these tasty fruits or better yet, have a date shake. Located at 59-111 Highway 111 in Thermal, the cafe/gift shop is surrounded by producing date palms. Go to www.oasisdate.com for more information.

Cabot's Indian Pueblo Museum

Located in Desert Hot Springs at 67-616 Desert View Drive, this is an unusual Hopi-style home built between 1941 and 1965 by the late Cabot Yerxa. Its 150 windows and 65 doors are a testament to the eccentricities of its builder. Go to www.cabotsmuseum.org for hours of operation and driving directions.

(continued on page 128)

The Coachella Valley Milkvetch,
Astragalus lentiginosus coachellae,
an officially endangered plant

Coachella Valley Museum & Cultural Center

Located in Indio at 82-616 Miles Avenue, the museum is dedicated to the history of the Coachella Valley with exhibitions and collections focusing on early settlers, Native Americans and historical events. This is an excellent place to learn about the life of early pioneers before air conditioning. Go to www.coachellavalleymuseum.org for hours and driving directions.

Palm Springs Historical Society

Two historic structures are the Society's center of activity. The McCallum Adobe, the oldest remaining building in Palm Springs, was constructed in 1884. Miss Cornelia's "Little House" was built by the city's first hotel proprietor in 1893. The structures are located at the Village Green Heritage Center, 221 S. Palm Canyon Dr. in downtown Palm Springs. Go to www.palmspringshistoricalsociety.com for more information.

General Patton Memorial Museum

Just off Interstate 10, at the Chiriaco Road turnoff east of Indio, is a museum dedicated to the great World War II hero. Patton spent many months in the immediate region preparing his troops for desert warfare. Telephone 760-227-3483 for information on driving directions and museum hours.

About the author . . .

James W. Cornett has lived and worked in the Coachella Valley since 1972. For nearly thirty years he was with the Palm Springs Desert Museum as its Director of Natural Sciences. He has written widely on desert subjects in his weekly newspaper column, popular magazines, scientific journals and more than two dozen books. Currently, his formal research activities are sponsored by the Joshua Tree National Park Association. When he is not giving public lectures or teaching University of California Extension classes, he enjoys birdwatching with his wife, Terry, in their backyard canyon at the edge of Palm Springs.